D1586459

FARMAGEDDON
ILLUSTRATED
The true cost of cheap meat

First published 2017 by
Bloomsbury Publishing Plc
50 Bedford Square, London WC1B 3DP
www.bloomsbury.com

Bloomsbury is a registered trademark of Bloomsbury Publishing Plc
Produced by Tall Tree Ltd

ISBN 978-1-4088-7346-5

A CIP catalogue for this book is available from the British Library.

Printed in China by Leo Paper Products, Heshan, Guangdong

1 3 5 7 9 10 8 6 4 2

Abbreviations

ha hectare
m^2 square metre
km^2 square kilometre
kg kilogramme
lb pound
$kcal$ kilocalorie

FARMAGEDDON ILLUSTRATED

The true cost of cheap meat

PHILIP LYMBERY

BLOOMSBURY

LONDON OXFORD NEW YORK NEW DELHI SYDNEY

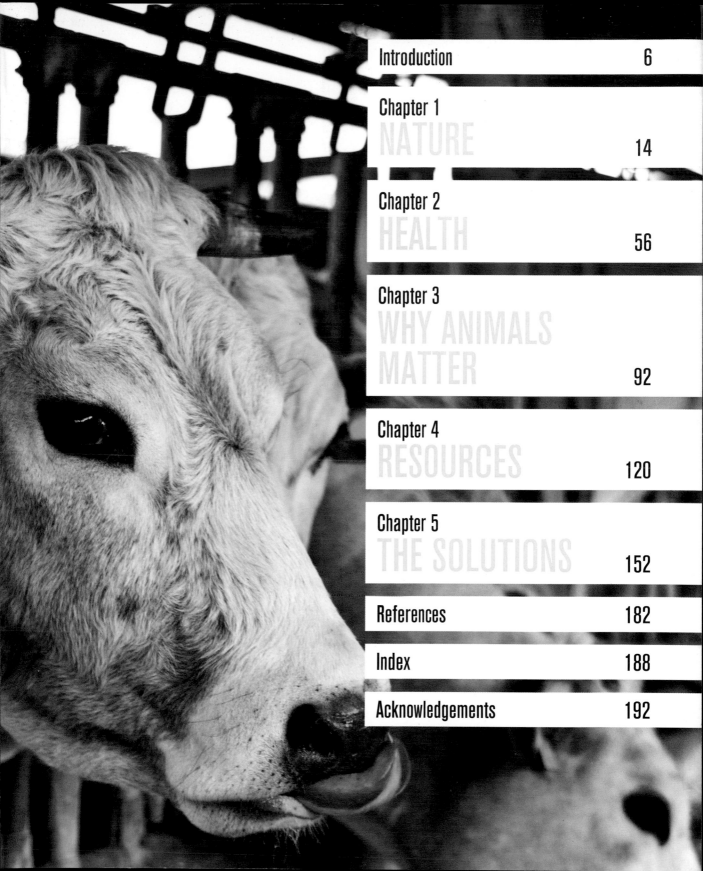

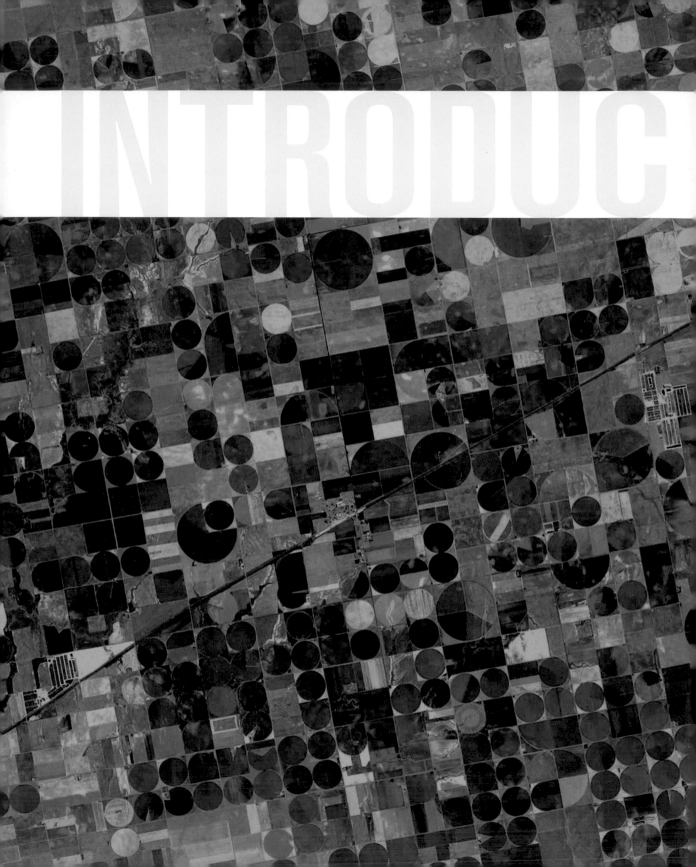

INTRODUC

TION

I live in the rural south of England, where pasture, hedgerows and wildlife are still very much part of the landscape. Yet under the guise of 'sustainable intensification', things are changing. Farm animals are slowly disappearing from fields and moving into cramped, airless hangars. Birds, bees and butterflies are vanishing. Governments have made it possible to buy chickens for £2, thinking they're doing everyone a favour. But the reality behind how 'cheap' meat is produced remains hidden.

People know things have changed, but it's easier to believe farms are like the Old MacDonald nursery rhyme – with chickens scratching round a yard, pigs snoozing in muddy pens and contented cows chewing the cud. It's a myth peddled to young children through picture books and songs, and reinforced through school trips to working farms that open to the public. But although fun, they're nothing like the average farm.

In fact, only eight per cent of English farms are now 'mixed' – rearing more than one type of animal and also growing crops.[1] Most specialise in just one product. These places would make a dismal day out for anyone, and would shock most schoolchildren.

'Bumblebees are major contributors to pollination of crops and wild flowers. We might face a pollination crisis.' Professor David Goulson

⬆ PESTICIDES
Modern GM crops are resistant to the superstrong pesticides sprayed on them, but the chemicals devastate populations of birds, bees and other insects.

FACTORY FARMS

Worldwide, some 70 billion farm animals are produced every year – two-thirds of them now factory-farmed. They live permanently indoors and are treated like production machines, pushed far beyond their natural limits. Since they're indoors with no access to grass or forage, their feed is transported to them, often across several continents.

Together, they eat a third of the world's cereal harvest,[2] 90 per cent of its soya meal and up to 30 per cent of the global fish catch[3] – precious resources that could be fed to billions of hungry people.[4]

It's a business that depends on vast quantities of antibiotics – half of all those used in the world[5] – and has created breeding grounds for new antibiotic-resistant 'superbugs'. It also eats up precious natural resources like oil, water and land.

A discovery made during the Second World War sowed the seeds of factory farming. While developing chemical weapons, German scientists discovered how to mass-produce organophosphate nerve agents which could be used as pesticides. Intensive farming practices were encouraged through new laws designed to combat food shortages suffered during the war years. At the same time, corn production expanded massively, making it a cheap source of animal feed.

The old patchworks of mixed farms were replaced by monocultures – farms specialising in a single crop or animal. Farming in tune with nature was no longer necessary. Chemical pesticides and fertilisers were now a quick fix for flagging soil. Factory farming was born.

I set out on a journey to see for myself the reality behind the marketing gloss of 'cheap' meat. I wanted to find out how things had changed, what's happened to our food and what it's doing to us. Over three years, I travelled with *The Sunday Times*' political editor, Isabel Oakeshott, to explore the complex web of farming, fishing, industrial production and international trade that affects the food on our plate. It was a journey that took me across continents, from the California haze to the bright lights of Shanghai, from South America's Pacific coast and rainforests to the beaches of Brittany.

This book looks at the consequences of putting profits before people and asks whether the Farmageddon scenario – the death of our countryside, new diseases and billions of starving people – is inevitable. It sheds light on what they don't want you to know and asks: could there be a better way?

Philip Lymbery

↑ CROSSING CONTINENTS
The author travelled to Peru to look at the impact of the fishing industry on the local people and environment.

INTENSIVE FARM
Each shed in this factory farm in the Peruvian desert contains many thousands of meat chickens.

CHAPTER 1
NATURE

INTRODUCTION

Back in 1962, Rachel Carson's book *Silent Spring* first raised the alarm over the effects of pesticides on our countryside.[1] The book's key message – about the dangers of pesticides and intensive farming – has largely been ignored in the 50 years since.

Pesticides and genetically modified (GM) crops play a major part in agriculture's new industrialised approach. Green pastures are being ploughed up and replaced with fields full of crops – but often not to feed to people. Instead, they are destined to feed farm animals taken off the fields and now living indoors in factory farms. These crops are often GM so they're super-resilient to the pesticides that kill most other plants around. But it's the loss of these other plants that is making birds, bees and butterflies simply vanish from the countryside.

Factory farms create another problem – muck. It used to fertilise pastures where it fell. Now, with animals indoors, it piles up, creating the headache of how to get rid of it. Because there's so much of it, this foul excrement often ends up polluting lakes, rivers and seas, seriously damaging the health of people living nearby.

The trend for ever-bigger factory farms is well established in the US, but now British businesses are trying to get in on the act. In 2009, plans for an 8,000-cow mega-dairy in Nocton, Lincolnshire, were scrapped after the Environment Agency complained about the risk of water pollution. The battle of Nocton, against the mega-dairy, may have been won, but the war against mega factory farms has just begun. It threatens to take the countryside, its wildlife and people to breaking point.

'The three biggest problems with pollution are agriculture, agriculture and agriculture.'

Betsy Nicholas

SPEED LIMIT 50

↑ SPREADING MUCK
The manure produced by factory farms is sprayed as fertiliser. Much of it leaks into the water supply.

CALIFORNIA GIRLS
A vision of the future?

California: according to the Beach Boys, home to the cutest girls in the world. Since the 60s, the state's population of females has soared, but the new arrivals share none of the glowing health and athletic physiques of the women in the song. Their purpose is to churn out supernatural quantities of milk before being turned into hamburgers.

A staggering 1.75 million dairy cows are reared in California,[2] crammed into barren pens on tiny patches of land and making a mockery of the vast potential space in this part of America. They pump out nearly 6 billion dollars' worth of milk every year,[3] and as much waste in the form of dung and urine as 90 million people.[4]

Instead of grazing on grass, the cows are confined to towering open-sided shelters, which stand in pens thick with mud and manure. The cows move very little as they are so heavy with milk, their udders the size of beach balls.

Between feeding and milking, there's very little for the animals to do but wait. Through a combination of selective breeding, concentrated diets and growth hormones designed to maximise milk production, they are pushed so far beyond their natural limits that they survive for just two or three years of milking before being slaughtered.

Pollution caused by industrial livestock production

7 [He] 2s² 2p³
N
Melting pt. -210.00 ˚C
Boiling pt. -195.79 ˚C
14.007
Nitrogen

Fertilisers used to grow feed crops have high levels of nitrogen.

Crops absorb only

30–60%

of the nitrogen in synthetic fertilisers.

Pigs assimilate

30%

and poultry

Therefore

40–70%

is lost to water or the atmosphere.

of the nitrogen in feed. Most is excreted in

manure

Nitrogen-use inefficiency when feeding crops to animals

Concentrated feed given to industrial livestock has high levels of nitrogen.

Nitrogen in crops fed to EU livestock

11.8 million

Nitrogen in meat produced for human consumption by EU livestock

2.3 million

| 14 | 12 | 10 | 8 | 6 | 4 | 2 |

million tonnes per year

SOYA FEED
**Dairy cows feed on soya from long
troughs in California's mega-dairies.**

The dairies arrived in Central Valley in the 1990s, after being pushed out of the Los Angeles suburbs. Land on the city outskirts was becoming more valuable, and as the population expanded, farmers were finding it cost more and more to dispose of waste. Many sold up and moved to the sticks, where they quickly discovered there was not much to stop them doing as they pleased.

FIRST ALARM

At the time, agriculture was exempt from California's Clean Air Act. It was not until the late 1990s, when cousins George and James Borba applied to build two 14,000-cow operations on adjacent properties in Kern County, in effect creating a 28,000-cow dairy, that serious attention began to be paid to the potential environmental and health impacts. The sheer scale of the plans galvanised people who were already worried by what they saw going on in Central Valley. Campaigners forced the authorities to undertake a full environmental impact assessment and the results were so alarming that such assessments became standard procedure. It also called into question the exemption of agriculture from the Clean Air Act.

The law was finally changed in 2003, and now farmers are supposed to comply with tough air- and water-pollution regulations. In practice, there is evidence that many routinely flout the law while overstretched authorities turn a blind eye.

Today, the Central Valley produces an incredible annual harvest of fruit, nuts and vegetables despite having so little rainfall that it's technically classified as semi-desert. While it might sound like the Garden of

PENNED IN
The cows spend the whole of their short lives in feedlots. They never see a field.

WASTE DISPOSAL
The muck from the mega-dairies leaks into local waterways.

Eden, it isn't. No grass, trees or hedgerows grow, except in private gardens and the ruthlessly defined fields.

The massive output of fruit and veg is only possible thanks to a cocktail of chemicals and the plundering of the crystal-clear rivers that run down from the Sierra Nevada mountains. By diverting waterways and dousing the parched soil with fertilisers, insecticides, herbicides and fumigants, farming has drained the soil of so much natural matter it might as well be brown polystyrene.

Sometimes, clouds of chemicals can be seen hovering above the crops, resembling the pollution that hangs over big cities. The dairies have such bad emissions that the surrounding air quality can be worse than Los Angeles on a smoggy day.[5]

With such an abundance of fruit in the area, you might expect to see lots of birds, bees and butterflies. The reality is there are very few. A fifth of children in Central Valley are diagnosed with asthma (almost three times the national average paediatric rate), and this is linked at least in part to the mega-dairy industry. Nearly a third of the four-million-strong population of Central Valley are assessed as facing a high degree of environmental risk, both from toxic air and from water pollutants.

"Living near mega-dairies is dangerous," says local campaigner Tom Frantz. "We are looking at a potential health disaster. I can see a new strain of *E. coli*, some kind of plague, breaking out in Central Valley. That is the worst-case scenario. It sounds remote but my worry is that it's just around the corner. Nobody will care until it is too late."

There are ten dairies within 13 kilometres (8 miles) of Frantz's home. The first arrived in 1994, the rest since 2002. Officially they house 70,000 cows, but the real total is likely to be far higher as dairies only have to submit the number of cows being milked at any one time.

PLAGUE OF FLIES

According to Frantz, the first thing residents noticed when the mega-dairies arrived was an influx of flies. Hardest hit was a school just over a kilometre from the first mega-dairy to open. Teachers used to keep doors and windows open in summer because they had no air-conditioning. These days, that's out of the question.

"There were swarms of black flies in the classrooms," Frantz explains. "It was difficult for the kids to work with them buzzing around. That first year they used rolls of sticky tape to catch them. Later, they installed screens on all the windows and sealed the doors."

Water quality is a major concern for many small communities in Central Valley. Studies have shown a direct correlation between intensive dairy farming and contamination of water wells, especially with E. coli bacteria and nitrates." Boil notices – ordering residents not to drink water from their taps without boiling it first – are a way of life here now.

"Ground water here is heavily polluted, mainly with nitrates, but also arsenic. "
Maria Herrera

Maria Herrera, a mother of four, runs the Community Water Center in the city of Visalia in the heart of the San Joaquin Valley. She says: "Ground water here is heavily polluted, mainly with nitrates, though there are also concerns about arsenic, which is linked to some fertilisers and is sometimes used as an additive in cattle feed. The meetings we hold with residents are always packed. The dairy farmers and their lobbyists come along and deny it has anything to do with them, but the evidence proves otherwise."

Kevin Hamilton, a registered respiratory therapist, became so concerned about what he saw through his work on the medical front line in Fresno that he has become a committed activist against mega-dairies. He says: "We're talking about heart disease, birth defects and stunted lung development among children who spent a lot of time outside playing sport."

SQUEEZED TOGETHER
Between milking, the cows are left with little to do. They are kept in a way that prevents them from forming the social structure of a herd in a field.

cows belong in

Join our campaign against mega

MEGA-DAIRIES: A DISASTER FOR ANIMALS,

"We're talking about high blood pressure and increased risk of stroke," continues Hamilton. "We have the second-highest level of childhood asthma in the whole of the US. Fifteen years ago, I couldn't have said any of that with confidence. Now the evidence is overwhelming. It's terrifying."

It would be easy to blame the farmers, but it's not as if they are all making fortunes. Most seem to feel under siege by environmentalist activists and regulation on a scale they never anticipated when they abandoned their small farms on the outskirts of LA.

Mega-dairy farming is a high-risk business exposed to global price hikes and price volatility. Evidence suggests they cannot weather recession as well as smaller pasture-based systems. The financial pressures involved in running a mega-dairy can be overwhelming. Chuck Cozzi, who owns a livestock market in Turlock, Stanislaus County, lost a close friend – who owned a large dairy – to suicide.

"You know, he had enough," he explains, in tears. "I think he shot himself, you know, leaving families behind. You know, kids. It's so sad. You know, I think maybe that guy

fields.org

dairies and factory farming
LOCAL PEOPLE AND THE ENVIRONMENT

was just so far in debt, he just gave up. Just got no more drive. I don't think anything could be that bad that somebody would want to do that, but it must have been for him."

It's a reminder that it's not just California's milk cows who are suffering from this bizarre perversion of farming. The dairy cattle are dying young, but so are some of the people who live and work with them.

In the land of the mega-dairy, humans, cattle and the environment are dancing to a grim tune of extraction and depletion. Each is just

⬆ **CAMPAIGN POSTER**
The mega-dairy model started in the US, but there have been attempts to export it to the EU. Campaigns to keep mega-dairies out based on environmental concerns have been successful so far.

an asset to be milked dry. These US mega-dairies – often 20 to 100 times the size of the average British dairy – provide us with a glimpse of how our planet might look if this style of farming becomes the norm around the globe, and they are now on the brink of migrating to the UK and other parts of the world.

DISAPPEARING SOILS

‘Man's continued existence is completely dependent upon six inches of topsoil and the fact that it rains.’ **Confucius**

45% of European soils face problems of soil quality, showing low levels of organic matter.

Soil biodiversity is under threat in **56% of EU territory**

Intensive agriculture is a key factor in loss of soil biodiversity.

Brazil loses **55 million** tonnes of topsoil every year due to erosion from soya production.

The UK may only have **100** harvests left because intensive farming has depleted the soil of the nutrients needed to grow crops.

The world has lost a **third** of its arable land due to erosion or pollution in the past 40 years.

Globally, approximately **33%** of soils are facing moderate to severe degradation.

Erosion rates
are on average 30–40 tonnes per ha per year in Asia, Africa and South America, and 17 tonnes per ha per year in Europe and North America (this means that soil is being lost at 17–40 times the rate at which new soil is formed).

17 tonnes

17 tonnes

30-40 tonnes

30-40 tonnes

30-40 tonnes

30-40 tonnes

The average rate of soil formation on cropland is around **1 tonne** per ha per year. It takes up to 1,000 years to form 2.5 cm (1 in) of soil.

NO-FLY ZONE UK

The creation of wildlife deserts across the UK

Ten million farmland birds disappeared from the British countryside between 1979 and the end of the last century. Not only birds vanished – farm animals did too. The two are linked. About 50 years ago, pigs, poultry and cows began vanishing from British fields to be reared indoors.

Mixed farms, where crops and animals were rotated around hedge-lined fields, started to go. Hedges vanished – 100,000 kilometres (60,000 miles) of them between 1980 and 1994 – and farmers began growing just one type of crop. More and more chemicals were being used to fertilise soil and wipe out pests.

In his book, *State of the Nation's Birds*, the late Chris Mead, of the British Trust for

Ornithology (BTO), spoke of sharp declines in once-common British farmland birds – skylarks, turtle doves and peewits. He described intensive farming as creating a "wildlife desert".

A devastating loss of seeds occurred on arable land: just a tenth remains of what birds could choose from 50 years ago, mainly because chemical herbicides have wiped out the

↑ **PEEWIT**
Once widespread, the peewit, or lapwing, thrived on mixed farms.

↑ **SKYLARK**
This small bird favours open country, where it feeds on seeds and insects.

seeding plants birds feed on. According to the BTO, the last 40 years have seen the population of tree sparrows crash by 97 per cent; grey partridges by 90 per cent; turtle doves by 89 per cent; corn buntings by 86 per cent; skylarks by 61 per cent; and yellowhammers by 56 per cent.[7]

UK government figures confirm what's been going on, with Britain's farmland bird numbers in free-fall, hitting an all-time low. They reveal a massive drop between 1976 and the late 1980s[8] – when UK farmers were switching to intensive systems. And the decline continues: in both the UK and Europe, farmland birds have declined more than those found in other habitats like woodlands or wetlands.[9]

Farming and wildlife can, and often do, go hand in hand. One example of this is Bickley

Hall Farm in Cheshire, where lifelong farmer Richard Owen has 140 hectares (350 acres) of permanent pasture and hay meadows grazed by cattle and sheep.

"Farmers are often blinkered by what they're doing and the amount of money they owe," he says. "Once they try this system of farming without inputs, with different breeds of livestock, that brings communities and people back to the farm, they become enthusiastic." At Bickley Hall Farm, the hedges are full of life and you can even find tree sparrows: rare in Britain nowadays, but doing well on this mixed farm.

The fall in bird numbers has slowed in recent decades, but the war against our wildlife continues – every day in every country where swathes of land are given over to single types of chemical-soaked crops.

DEAD ZONES
Agricultural pollution has fed algal blooms in Chesapeake Bay. The blooms reduce oxygen levels, killing fish, clams and worms.

WATERWAYS CHOKED BY CHICKENS
The death of Chesapeake Bay

Chesapeake Bay in the US was once the scene of violent clashes over shellfish. The bay is well known for its oysters, but they're in trouble – down to less than one per cent of their former numbers thanks to over-harvesting and pollution. Today, The Bay's biggest battle is against chickens and the muck they produce.

"The three biggest problems with pollution in this area are agriculture, agriculture and agriculture," says Betsy Nicholas, executive director of Chesapeake Waterkeepers. "There are huge problems here with algal blooms, times when there are complete dead spots in the Bay."

Farmland covers a quarter of the Chesapeake Bay watershed and is the largest single source of pollution, much of it relating to poultry manure.[10] Betsy explains that local people are worried about tackling the pollution problem in case they are seen as attacking family farmers.

"But factory farms are a different matter," she says. "It's really become an industrial pollution source and really needs to be addressed as such." The area used to be big on producing timber, orchard fruit and vegetables like tomatoes and cucumber. Now the dominant 'crop' is chickens. Many of

The pungent smell of chicken manure is a familiar part of spring.
Kathy Phillips

what were once family-owned farms have been taken over by big business. Factory farming here is on a colossal scale. One farm can have as many as thirty huge hangar-like sheds covering up to 2,000 square metres (21,000 square feet), each crammed with thousands of birds that remain indoors for their entire lives. They churn out millions of chickens every year, yet if you visit the area, you're unlikely to see a single bird.

Where there were once small family-owned farms producing crops, there are now huge factory farms. They cover as much ground as a small shopping centre. The difference is that a proposal for a shopping mall goes through lots of planning processes, whereas farms are treated more lightly. Often the first you know about it is when they're already being built.

"There are now nearly as many chickens being produced in the states surrounding the Chesapeake Bay as there were across the entire country 60 years ago," says Bob Martin, senior policy adviser at John Hopkins University, Boston. "The pollution from the chicken factory farms is causing a reduction in natural seabed grasses, making it harder for the oysters to grow," he continues. "Local crabs are growing bigger and eating the oysters … things are out of balance."

Bob sees industrial agriculture as the biggest threat to the environment and public health in this area. "I enjoy eating meat," he confesses, "but there needs to be some common sense in what we do."

SPRAYING MUCK

Corn and soya grow in nearby fields and are used to feed the chickens. A foul stench fills the air in spring as manure from chickens is sprayed on the fields to fertilise those crops.

Muck is also piled in huge muck mountains or dumped in festering lagoons as big as Olympic-sized swimming pools. There is so much muck that lagoons often leak or seep into nearby waterways, causing serious damage to both water and wildlife.

Clean-water campaigner Kathy Phillips came to Maryland for the beach life in the 1970s. She now enforces federal laws to protect the local coastline. "The pungent smell of chicken manure is a familiar part of spring here," she says. "CAFOs [Concentrated Animal Feeding Operations, or factory farms] are everywhere in this area."

When asked if she buys cheap chicken, Kathy says: "No I don't. Buying chicken in the supermarket is cheap food all right, but what people don't see is the hidden cost. It's their taxpayers' money that's being used to

HIDDEN COST
In 2006, it was estimated that it would cost $15 billion to clean up Chesapeake Bay.

put up manure sheds, and vegetative buffers around poultry houses to catch the dust and ammonia emissions. State money also goes into moving the manure out of the watershed."

Chesapeake Bay has been placed on an unprecedented 'pollution diet'. Several states and the Environmental Protection Agency (EPA) are working hard to cut the nutrient flow into the water. Yet despite their best efforts, they've failed to meet targets and more than half of the streams in the Chesapeake watershed are described as 'poor' or 'very poor'. Gone are the snails, insects and other waterborne wildlife essential for a healthy aquatic environment.[11]

THE ETERNAL SUN DANCERS
A natural wonder under threat

Catching butterflies sounds like a cruel and old-fashioned hobby, but in the United States now it's quite the thing. These days, the aim is not to kill butterflies and turn them into works of art but to attach a tiny tag to their wings then set them free. Thousands of proud Americans now own butterfly nets and take part in an extraordinary conservation programme involving people of all ages, called Monarch Watch.

The monarch is an emblematic species in North America, the only type of butterfly that the average American knows. Every year it's the star of one of nature's most spectacular shows – a migration that takes it thousands of kilometres from the northern states of the US to Mexico and back again.

On the way, these delicate creatures face fierce challenges. At around 12 km/h (7 mph), they make their way past motorways, chimney stacks, pylons and power stations belching out noxious gases. Millions die en route, blown apart in the slipstream of aeroplanes, splatted on car windscreens, drowned in puddles and birdbaths, burnt in bonfires and trapped in webs. Despite these hazards, millions of others make it.

Starting on the same date each year – the autumn equinox – tens of millions of these butterflies migrate from the US and Canada to one small area, the pine and fir forests of the Neovolcanic Plateau, some 390 kilometres (240 miles) from Mexico City. The following March, other butterflies head off on a return journey. Each monarch travelling in either direction is a member of a new generation, making the journey for the first time. They are guided by an internal navigation system, but quite how they do it remains a mystery.

The roosting site is extraordinary. Great clumps of butterflies coat entire tree trunks and branches in what looks like jagged orange and black fur. Most of the butterflies rest, wings folded, on the pines, but others swirl between the trees in dizzying masses. People who have witnessed it say the biggest surprise is the noise of all the beating wings – it sounds like rushing water.

DECLINING NUMBERS

Monarchs are the only butterflies to migrate like this. Yet these superb migrants are under threat: the number of monarchs volunteers can find to tag has dropped by almost a third in the last eight years. Their current population is the lowest on record. The finger points at pesticides.

At the heart of the problem is a pink-flowered plant called milkweed. Monarchs love it and won't lay their eggs on anything else. But it's also a weed, and farmers don't like it. The use of genetically modified crops – which are resistant to super-strength herbicides that destroy virtually everything else – has all but wiped out the plant in many areas.

"Monarchs are dependent on milkweed plants, and milkweed plants are weeds," explains Dr Orley 'Chip' Taylor, an insect

ecologist at the University of Kansas. Milkweed used to be controlled more naturally by good farming practices: "There would be a small number of plants per acre that would survive in corn and soybean fields … But in 1996 they developed a herbicide-tolerant soybean, and then they came up with herbicide-tolerant corn … Eventually what they did was eliminate the milkweed."

Experts fear that monarch numbers are so low that the migration will either disappear

NATURAL WONDER
Millions of monarch butterflies swarm around their roosting site in Mexico, covering entire trees.

altogether or no longer bear any resemblance to the magnificent spectacle it is today. "Here we have one of the world's most magnificent biological phenomena and to lose this would be absolutely drastic," laments Dr Taylor. "I mean, this is one of the most spectacular events that happens on the planet."

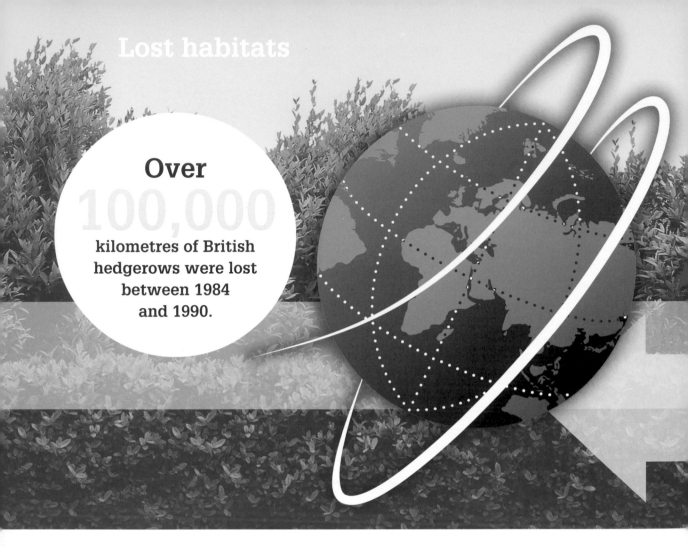

Lost habitats

Over 100,000 kilometres of British hedgerows were lost between 1984 and 1990.

It's not just monarch butterflies that are suffering as a result of intensive farming. In the UK, farmland is the main habitat for three-quarters of butterflies, and conservation groups say that, typically, they now survive only on relatively small areas. Wild-flower meadows, where no fertilisers or pesticides are used, are vital as they can provide habitats for up to 25 different species. A report by the UK Butterfly Monitoring Scheme, which tracks the wellbeing of 54 types of butterfly, found that 30 had declined over the last ten years. A total of 21 species declined by more than a quarter, while 12 species declined by half or more. Victims include iconic British species such as the red admiral, painted lady and holly blue.

But there is hope. In 1979, the large blue butterfly was declared extinct in Britain. Intensive farming was blamed for destroying much of its habitat. It was particularly regrettable because the species is threatened globally. The conservationists refused to give

At the end of the 20th century, the populations of **67%** of 333 farmland species (broadleaved plants, butterflies, bumblebees, birds and mammals) were declining due to agricultural practices.

That's the equivalent distance to going around the world twice

up, however – they reintroduced the large blue to the British countryside in what was described as 'the world's largest-scale, longest-running conservation project involving an insect'. Twenty-five years later, its delicate pastel-blue wings can be seen again over more than 30 former sites.

As for the monarch, there are large non-migratory populations in other countries, so for now their future seems secure. Yet its decline is a symptom of a far bigger problem.

"It's not that monarchs are important in themselves, it is that monarchs are symbolic of how we're doing. If we can't support the monarch butterflies, it means we're not supporting a lot of things, because monarch butterflies share habitats with virtually all pollinators in the US and they share habitats with a lot of small mammals and birds." Butterflies, likes birds and bees, are the hallmarks of a healthy environment. Their plight is part of the complex web of life that underpins food and farming.

↓ FUSSY EATER
Monarch butterfly caterpillars feed only on milkweed, which is being wiped out by modern pesticides.

BUZZED OFF
Bees under threat

Both wild bumblebees and domestic honeybees are under severe threat. In the UK, out of 24 types of bumblebee, two species have already gone extinct in the last 70 years. Six are considered seriously endangered,[12] and half the rest are at risk.[13] The British Beekeepers' Association fears the UK could lose all its bees within the next decade.[14] In the US, several recently common species have disappeared. It's the same in other parts of the world.

The potential implications of the decline are catastrophic. Most fruit and vegetable crops depend on pollination by bees. The future of around a third of global agriculture is at stake.

There is still some debate over the cause of the collapse of the bee population, but most experts think it's due to agricultural intensification, particularly the use of chemical pesticides.[15] Artificial fertilisers mean there's no need to rotate crops, and herbicides have wiped out most of the wild plants bees used to forage on. World bee expert, Professor David Goulson, says: "Bumblebees are major contributors to pollination of crops and wild flowers throughout the temperate northern hemisphere. Many species have declined, contributing to fears that we might face a pollination crisis.

"In Europe, the primary driver is thought to be habitat loss and other changes associated with intensive farming. In the Americas, declines of some species are likely to be due to impacts of non-native diseases."[16]

Governments have been slow to wake up to the problem. In 2007, the US House of Representatives held an emergency hearing on the issue to discuss it and set aside $5 million for honeybee research, but the amount was later slashed by half.[17] Farmers can't afford to sit and wait, so they're taking their own desperate measures – hiring in commercially reared bees to pollinate their crops.

Industrialised pollination is now commonplace. In California, commercially reared bees are brought in by the truckload from as far afield as Australia to support the state's almond industry.[18] Every year, some 3,000 trucks drive across the United States carrying around 40 billion bees to California's Central Valley, home to more than 60 million almond trees.

California growers now spend an astounding $250 million a year on bees.[19] Many beekeepers now make more money from pollination services than from honey.

COMMERCIAL HIVES
About half of all the commercially bred bees in the United States are brought to California each year to pollinate the state's almond trees.

43

Charges for bee services are soaring – prices have tripled since 2004. Dr Parthiba Basu, an ecologist at the University of Calcutta, has linked falling bee populations to poorer crop yields. He made the connection by accident while investigating whether farmers in India could viably turn their backs on intensification. The answer was a resounding yes. Those that switched to mixed farming – rearing several types of livestock and crops in the same place – made more money.

They also found that crops that were highly dependent on bees for pollination were not doing as well as expected. "It was particularly striking in Bengal, where they grow a lot of a vegetable known as painting gold," says Basu. "It's a highly pollinator-dependent crop. They have been hand-pollinating for a while now, because there are just not enough bees left." Basu is now studying bee problems in other developing countries, and he believes agricultural intensification is to blame.

DEADLY SPRAY
Californian almond groves are routinely sprayed with pesticides.

BEE BUSINESS
It takes four colonies of bees to pollinate one hectare (2.5 acres) of almond trees.

He says: "I had hoped that pollinator loss would not be nearly as serious in developing countries as it is in the West, but that does not seem to be the case. It is very sad. It is going to take a lot of effort to turn this around, but unfortunately the developing world is going down the opposite route right now, embracing Western-style intensification."

Achim Steiner, director of the UN's Environment Programme, believes that we are labouring under the illusion that we have the technological prowess to be 'independent of nature'. As he puts it, bees "underline the reality that we are more, not less dependent on nature's services".

Toxic pollen

Around 30% of human food crops are pollinated by bees, much of this by wild bees, and the proportion of pollinator-dependent crops is increasing worldwide. Both wild bumblebees and domestic honeybees are endangered by the practices of modern agriculture.

US honeybee colonies have been shown to be contaminated by over 100 different pesticides.

31 different pesticides have been detected in a single pollen sample.

 different pesticides have been detected in a single wax sample. **39**

60% of pollen and wax samples contained at least one systemic pesticide.

PLAYING SCALES
Big aquaculture

Fish farms are the forgotten factory farms under the water. Around 100 billion farmed fish are produced globally every year – 30 billion more than all the chickens, cows, pigs and other farm animals reared worldwide.[20] It's one of the fastest-growing sectors of intensive farming.

As natural fish stocks plummet worldwide, people often assume that fish farming is a happy solution. The reality is these farms plunder the oceans for smaller fish to feed the salmon, trout and other fish they produce. It takes between three[21] and five tonnes[22] of small fish to produce one tonne of farmed fish like trout or salmon. Supplies of wild fish to feed the captives are now in grave danger of running out: over half of all wild fish are already 'fully exploited' and almost a third have been overexploited.[23]

As in other factory farms, farmed fish are crammed into small spaces – up to 50,000 salmon in a single sea cage. Often suffering from blinding cataracts, fin and tail injuries, and infested with parasites, they are forced to fight for space and oxygen. Each salmon gets the equivalent of a bathtub of water to swim around in. Packed so tightly, they end up swimming round in circles. Fins and tails rub sore as the fish press against each other and the sides of the cage. Trout get even less space – about the same as 27 trout, each the length of a ruler, being squashed

FISHY BUSINESS
Half of all the fish we eat today comes from farms like this one. In 1970, just 5 per cent of our fish was farmed.

together in a single bathtub of water. This crowding and lack of space makes them more likely to become ill. In recent years there have been many outbreaks of illness, killing millions of fish. Death rates among farmed salmon in sea cages can be so high they would cause serious alarm in other farmed animals.

SPREADING LICE

Fish farms pose a massive threat to wild fish, as they are hotbeds of sea lice.[24] These parasites, which look like tiny tadpoles, are a major problem.[25] They latch on to fish and eat away at their skin and scales. It can be so relentless around the head that the bone of the living fish's skull can be seen – a condition known as the 'death crown'. Young wild fish can cope with brief exposure to sea lice, but when they swim past salmon farms they can suffer prolonged exposure, which can kill them. A study carried out in British Colombia found that if outbreaks of sea-lice infestation continues, "local extinction of wild salmon is certain, and a 99 per cent collapse in pink salmon population … is expected in four salmon generations".[26]

Wild fish are genetically adapted to their natural surroundings. Even moving a salmon to a different river can make it less likely to survive. This suggests that mixing wild and farmed salmon genes makes them weaker.

Wild fish are also under threat from escaped farmed fish. This happens regularly thanks to careless handling or damaged nets and cages. Tens of thousands a year can escape. The fast-growing farmed salmon compete with wild fish for food and have even been known to prey on young wild salmon.[27] As farmed and wild fish interbreed, their offspring are less likely to survive.

The impact of fish farming on the wild populations might be offset if farmed fish were as tasty and wholesome as the organic alternative. But they're not. Farmed salmon and trout, for example, contain more fat than wild ones. Figures show that, for the same amount of protein, farmed Atlantic salmon has twice as much fat as its wild cousin, and farmed rainbow trout is up to 79 per cent fattier.[28]

Extra fat isn't the only health concern with farmed fish – they may also contain worrying levels of contaminants. One study found that concentrations of chemicals were "significantly higher" in farmed fish samples. Its conclusion was damning: "Consumption of farmed Atlantic salmon may pose risks that detract from the beneficial effects of fish consumption."[29]

‘ It's a true saying that when you buy a Scottish salmon, you pay for bullets to shoot seals. ’
John Robbins

Swimming space

Many of the 100 billion fish that are farmed globally each year are crowded together in pens and sea cages, where they are unable to exhibit their natural behaviour.

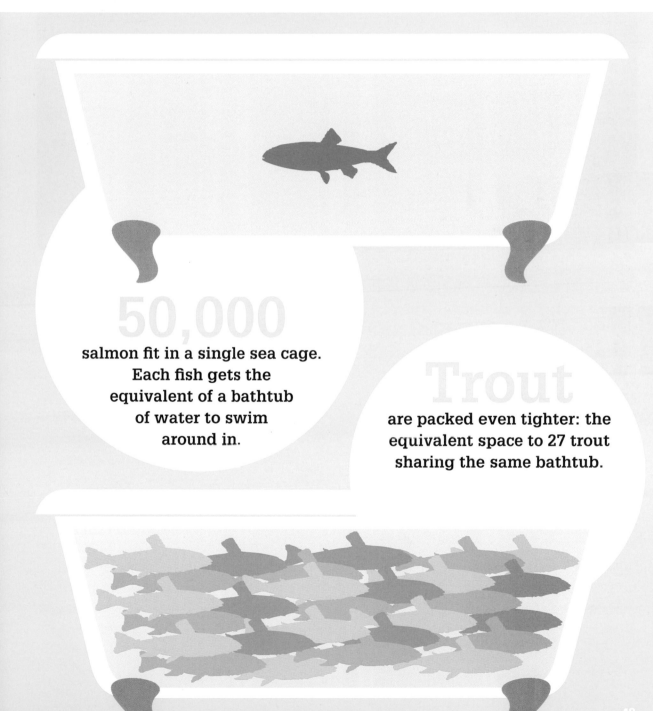

50,000 salmon fit in a single sea cage. Each fish gets the equivalent of a bathtub of water to swim around in.

Trout are packed even tighter: the equivalent space to 27 trout sharing the same bathtub.

49

As well as being treated with chemicals to combat sea lice, farmed salmon and trout are fed chemicals to make their flesh turn pink – something that happens naturally in wild fish. Other marine life can also be badly affected by fish farming. Huge numbers of fish in one place attract seals, birds, otters and other wildlife. Many farms shoot them to protect their stock. By law, shooting is now supposed to be a last resort, but evidence suggests some farmers still see it as the cheapest and easiest solution.

There are kinder alternatives – anti-predator nets. But John Robins, an animal welfare campaigner in Scotland, says many salmon farmers don't bother to invest in them. "You'd have to be shooting seals 24/7 to fully protect the fish. You really do need

↑ FISH DEATHS
Interbreeding between wild and farmed fish can have catastrophic effects.

anti-predator nets," Robins explains. The problem is that these nets are expensive, and Robins has discovered that 80 per cent of Scottish fish farms don't even have them. "That does not make shooting a last resort, it makes it very much a first or second resort."

The Scottish government claims the number of seals being shot has dropped dramatically in recent years. But it goes on in remote areas, with the government relying on the honesty of shooters to provide figures of how many they kill. Robins believes: "It's a true saying that when you buy a Scottish salmon, you pay for bullets to shoot seals."

ANIMAL CARE
What ever happened to the vet?

The traditional image of a vet trundling about the countryside tending to sick animals started with James Herriot – the hero of the best-selling book and TV series *All Creatures Great and Small*. Yet there is a darker side to the veterinary profession, which is rarely seen by those outside the industry. A growing army of vets do work not in small town surgeries microchipping dogs and patching up injured cats, but in dimly lit sheds on industrialised farms and in abattoirs.

These vets prop up the factory-farming system by keeping animals alive long enough for profitable slaughter or to churn out enough milk or eggs to justify their existence. Few vets start their careers intending to work in such places, but slaughterhouse jobs have regular hours and involve more observation and inspection than hands-on intervention. They therefore suit older vets or those with bad backs and other health issues. They also provide a genuine chance to do good by minimising the pain and suffering of tens of thousands of animals in the hours before they are slaughtered.

Live transport

Animals are transported from farms in the EU in overcrowded conditions. If they survive the journey, they often face a cruel death at their final destination.

Every year more than 3 million live cattle and sheep are sent from the EU for slaughter in Turkey, the Middle East and North Africa.

That is 8,600 live animals per day.

Jean-Claude Latife – a French vet whose real name cannot be revealed for legal reasons – spent nine years working in UK abattoirs and alleges that they often employ untrained workers who turn up for work drunk or take drugs at work. He claims vets are frequently pressurised or intimidated into turning a blind eye when they witness staff mistreating animals.

It's a far cry from what Latife imagined when he first moved to the UK. As a small-animal vet in France, he had developed various medical problems and was advised by doctors to look for a less stressful job. He and his family had fallen in love with the English countryside on holidays, so when a position came up in a slaughterhouse in one of the most beautiful parts of the country, he jumped at the chance.

CUTTING CORNERS

By law, abattoirs in the UK must have a vet on-site. Their job involves witnessing animals arriving and being unloaded, checking they are in a fit state to be transported, and ensuring that conditions on the lorry are acceptable. They also keep an eye on how animals are handled before they are killed and whether they are properly stunned before slaughter. They have the power to stop the production line if they are unhappy about an issue of animal welfare, contamination or hygiene.

Latife grew quickly hardened to being insulted by workers frustrated when he questioned procedures or caused delays to production to safeguard animal welfare. However, when a slaughterman who had previously served six

During transport
The long journeys are extremely stressful. The trucks and ships carrying the animals are often overcrowded, with inadequate ventilation and water supplies. By journey's end many of the animals are exhausted and dehydrated.

At slaughter
Investigations show EU cattle being slaughtered outside butchers' shops. Cattle often have their leg tendons severed in order to control them. Some EU cattle are placed in boxes that turn them onto their backs; immediately after throat-cutting they are ejected from the box while still conscious and fall onto the bodies of other dying cows.

months behind bars threatened to kill him, he knew it was time to quit. "One day, a guy I normally got on well with went crazy," says Latife. "I knew he took drugs and that day he took something during his tea break. When he came back, I noticed there was not enough space between two carcasses, which raises a risk of cross-contamination. It meant stopping the line. When I told him we had to stop the line he got very mad … He started insulting me and threatening my wife and children. At the same time, another guy came at me with a knife. It was terrifying."

Latife's account of UK slaughterhouse workers, and the way some encourage staff to cut corners in animal welfare by paying them 'by kill' rather than hourly rates, is not an isolated case. Recent undercover filming in nine British slaughterhouses revealed evidence of cruelty and lawbreaking in most of them – including animals being kicked, beaten and burned with cigarettes. An exposé of 25 abattoirs in France uncovered rough handling and poor stunning practices.

Given their professional status, vets are in a unique position to make a positive difference. Yet Latife's experience and evidence from investigations suggests many are intimidated into keeping their mouths shut over routine abuses and only intervene over blatant breaches.

Their role isn't only confined to the end of a factory-farmed animal's life. Vets play a vital role at every stage, often treating and preventing disease, supporting a system where suffering is inbuilt. The harder farmers push animals beyond their natural

53

limit, and the more closely animals are confined, the more farms have to rely on vets to keep herds alive. Their weapon of choice is antibiotics.

According to Dil Peeling, a vet who qualified in the UK but spent most of his career in developing countries, "A vet's worth is now measured by his or her ability to deliver on production and animal health – not welfare. It's difficult to persuade vets who have invested so much of their careers propping up intensive farming to turn their backs on such systems … As far as they're concerned, this is how things have always been done." Peeling, who now works for The Brooke, an international horse welfare organisation, believes the industry is geared towards rewarding vets who focus on farm animal health in terms of how it relates to production, rather than seeing the welfare of the animals as an end in itself.

The vast majority of vets who work on farms genuinely care about the animals they work with and are upset when they witness suffering. They would alert the authorities to flagrant welfare issues. Alistair Hayton is one of the UK's leading cattle vets and works on a variety of dairy farms in Somerset, some with over 1,000 cows. He regularly visits an organic farm and a large-scale dairy farm and says he is happy with conditions on both. "I am convinced there is no right or wrong system, only right and wrong management," says Hayton. "Big is not necessarily bad. I have seen some very bad practice on small farms. The key is the individual farmer – how well they manage their farm, and whether their facilities are appropriate for the number of animals being kept."

However, Hayton acknowledges that intensive farming is highly fine-tuned and that some farmers are increasing the size of indoor-

reared herds without extending their barns. Some systems don't have much scope for decent welfare. Take battery cages where each hen only has enough space to stand. They spend their entire lives standing on bare wire and can't even flap their wings. The system is so intrinsically limited and restrictive that the birds are bound to suffer frustration and ill health. No amount of stockmanship will create decent welfare in those cages.

Supporters of factory farming claim battery hens wouldn't lay eggs if they weren't happy. Yet the truth is they're genetically programmed through selective breeding to lay about 300 eggs a year, and will do so whatever conditions they are kept in, so long as they have food and water. Governments will continue to have a role in shaping how things are done, be it through legislation or how they dole out public subsidies or incentives; but business can move far more

↑ CAGED IN
Caged hens live in such cramped conditions that suffering is inevitable.

quickly and decisively than government. While leading businesses offer the scope to be part of the solution and governments in Europe have banned some of the worst forms of farm animal cruelty, the sad truth is that industrial agriculture still has stranglehold in much of the developed world. And with industrial farming often comes an attitude that animals are mere units of production, a means to make a profit.

James Herriot once said he hoped to help people realise how "totally helpless animals are, how dependent on us, trusting as a child must that we will be kind and take care of their needs … [They] are an obligation put on us, a responsibility we have no right to neglect, nor to violate by cruelty."[32]

CHAPTER 2
HEALTH

INTRODUCTION

Deadly diseases like cancer, diabetes and heart disease are far more prevalent than they were 70 years ago.

One in three people now get cancer. "This is not just down to people living longer or to better detection," says Dr Michael Antoniou, a molecular biologist with a wide-ranging interest in human health. "The incidence is going up due to environmental factors." He believes that farming is partly – even largely – to blame, thanks to its industrial approach and vast use of pesticides and fertilisers. These cause a toxic cocktail effect out in the countryside.

Meanwhile, there's an obesity crisis – epidemic even in rich countries. Some blame diets high in saturated fats, such as cheap meat, as well as a lack of exercise. At the same time, more and more antibiotics are being given to farm animals to boost growth and prevent diseases. Strong evidence links this to the arrival of superbugs like MRSA. In China, where half the world's pigs live, unchecked antibiotic use and pollution by mega-farms are widespread and on such a big scale that it's putting global public health at serious risk.

One of factory farming's filthiest secrets is the fishmeal factories that export animal feed to factory farms around the world. These factories are spitting out pollution in places like Peru, causing skin lesions, asthma and diarrhoea in children.

Are factory farms making us sick?

'By confining billions of animals on factory farms, we have created a worldwide natural laboratory for the rapid development of a deadly virus.'

Dr Aysha Akhtar

↑ SPREADING DISEASE
Intensive broiler farms are the perfect breeding ground for new viruses.

BUGS AND DRUGS
The misuse of antibiotics

When the idea of farmers feeding antibiotics to their animals was debated in the UK parliament in 1953, it sounded great. MPs were told that feeding tiny amounts of antibiotics to pigs could have a "most remarkable" effect on their growth. The then Health Minister said there could be "no adverse affect whatsoever upon human beings".

But a few politicians were not convinced. They warned that the consequences could be disastrous. Hugh Linstead, MP for Putney, told Parliament: "We have not been doing it long enough, I feel, to know what effect it will produce in the long term in herds and on meat, and indeed, on human beings who eat the meat …"

He wasn't the only one to be concerned. Dr Barnett Stross, MP for Stoke-on-Trent Central, said: "If pigs are fed in this way, new types of bacteria may evolve and thrive which are resistant to the penicillin that the pigs are eating regularly in their food … Should that arise, it would mean first that we should lose the benefits that we are now about to gain…if there be migration of the bacteria to humans we may find ourselves in trouble …"

Back then, Dr Stross's warning was greeted with guffaws.[1] Today, events have unfolded almost exactly as Linstead and Stross predicted. Antibiotics, the 'wonder drug' of the medicine cabinet, are now so widely used and abused – not only in human medicine but on farms, too – that they're losing their potency on people.

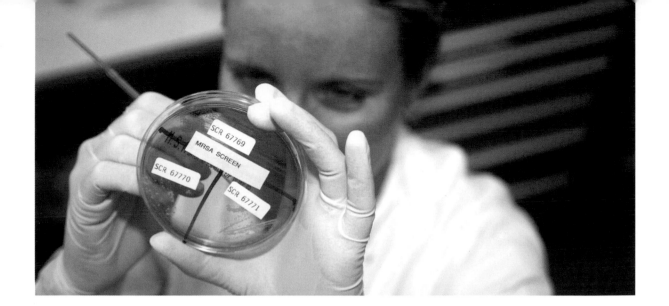

In 2008, the UK's then Chief Medical Officer, Liam Donaldson, warned that bacteria were becoming so resistant to antibiotics that: "In some diseases … the last line of defence has been reached."[2] On World Health Day in 2011, the Director General of the World Health Organisation (WHO), Dr Margaret Chan, warned of a "post-antibiotic era, in which many common infections will no longer have a cure and once again, kill unabated".[3] If that happens there will be no effective treatment for killer diseases like typhoid, tuberculosis, pneumonia, meningitis, tetanus, diphtheria, syphilis and gonorrhoea.

Antibiotic misuse in people is a big part of the problem but their overuse on farms is helping to bring this medical Armageddon even closer. Half the world's antibiotics are fed to farm animals, largely to ward off the diseases of factory farming. Every dose of antibiotics given to a person or animal is a chance for resistant bacteria to develop.

MUTATING BUG
Scientists are finding new strains of MRSA, some of which are resistant to a range of antimicrobials.

Antibiotic growth promoters are now banned in the EU – though some get round the law by using low doses to ward off disease whilst also boosting growth. The European Medicines Agency described factory farms as places that provide "favourable conditions for selection, spread and persistence of antimicrobial-resistant bacteria".[4]

Pathogens can be passed on through meat and to people who work with infected animals, through manure or even in airborne particles. It can then spread from person to person. Of the so-called superbugs, MRSA is the most notorious. Until a few years ago it was found mainly in hospitals but in many

A BETTER WAY
Free-range, organic farms rear animals safely without the routine use of antibiotics.

countries, the bug is now striking people who've had no contact with hospitals – so-called 'community acquired' outbreaks.[5] A previously unknown strain of MRSA found in pigs has begun spreading to people. The first infected were a Dutch baby girl and her parents, who were pig farmers. Now half of all Dutch pig farmers are thought to carry the strain – 760 times the average rate in the wider population.[6] Dutch scientists and government officials point the finger at intensive pig rearing and its use of antibiotics for the rise and rapid spread of farm-animal MRSA.

Scientists have found 15 cases of a completely new type of MRSA in milk from English dairy farms. It's already infecting people in England and Scotland – though not from drinking milk, as pasteurisation kills bacteria.[7] MRSA is just one threat to human health created by antibiotic use on factory farms. Rising numbers of food-poisoning cases are now antibiotic-resistant, which means catching these bugs can be life-threatening.

In 2009, there were almost 200,000 reported cases of food poisoning from the bug campylobacter in the EU and around 109,000 cases of salmonella.[8] But as most people don't bother to report food poisoning, the real total could be as many as two million. Meanwhile, organic farmers have shown it is entirely possible to raise healthy animals with minimal use of antibiotics.

ANTIBIOTIC DANGER

The health risks of confining animals

The numbers of infections made more difficult to treat by antibiotic resistance are expected to increase markedly over the next 20 years.

Without antibiotics, a bacterial blood infection could kill around 80,000 people.

Approximately 15% of patients having a total hip replacement would die of infection without antibiotics.

An estimated

45%

of antibiotics used
in the UK are used
on animals.

Approximately

two thirds

of all antibiotics used in
26 European countries
are used on farm animals.

Up to 70%

of medically important
antibiotics sold in the
US are given to food
animals, not people.

Alliance to Save our Antibiotics recommends

1. Reducing antibiotic use by 80% by **2025**

2. A ban on preventative mass medication in feed and water, except where disease has been found.

3. Collection of data on antimicrobial resistance among humans.

65

FLU, BUT NOT AS WE KNOW IT

The new bugs threatening a pandemic

The recent lethal outbreaks of bird and swine flu also have strong links to factory farming.

The H5N1 bird flu virus emerged at a time when the Far East poultry industry was expanding massively. It was first spotted in Hong Kong's live-bird markets and chicken farms in 1997, when six people died. From 2003, it spread across East Asia, exactly when the poultry population was growing, becoming more intensive. It then spread across Asia, the Middle East, Europe and Africa. By August 2011, 564 people had been infected, 330 of whom died – a death rate of almost 59 per cent.[9] Most people who caught it worked with chickens.

An outbreak of another strain of bird flu in the Netherlands in 2003 showed that these bugs can pass from poultry workers to other people.

This time, a vet died and tests found that 86 poultry workers and three family contacts had been infected with the disease.[10] The big fear is that every time someone is infected, the virus mutates further, which can make it far more contagious. This could lead to a global epidemic. Recently, scientists found that just a few mutations would allow H5N1 to become as infectious as seasonal flu.[11] An article in the *New Scientist* described the risk of a pandemic as "fact, not fiction", and *The Lancet* estimates that a flu pandemic could kill as many as 62 million people.[12]

We now know the long-distance transport of animals plays a major role in the spread of farm diseases. As well as causing suffering to

➤ LIVE BIRD MARKET
Fresh outbreaks of bird flu are still being found in East Asia. Live markets such as this one in Taiwan help it to spread.

➤ BIRDS AND BUGS
Intensive chicken farms provide the ideal conditions for bugs to spread and mutate.

the animals involved, it allows diseases to 'hitchhike' to new places and populations. Some have tried to blame wild birds for avian influenza (AI) and use it as an excuse to keep poultry in intensive indoor systems. But low-level avian flu is natural in wild birds, and it's only when the disease enters the overcrowded intensive farm that it mutates dangerously. Bird flu has been shown to spread along major road and rail routes rather than on the routes flown by migratory birds. While the overcrowded factory farms provide the ideal places for new aggressive strains of bird flu to appear, wild birds can then become infected too.

When H5N1 hit a Bernard Matthews turkcy farm in Suffolk in 2007, there was no evidence of a highly pathogenic flu in the wild bird population. Defra reported at the time that more than 4,000 wild birds had been tested over the previous six months and only 0.4 per cent had AI. None of those infected had the same dangerous type that was involved in the outbreak. The main source of the problem is the factory farming system itself.

HUMAN HEALTH

Saturated fats such as animal fats contain cholesterol at **much higher levels** than vegetable fats.

Diets high in saturated fat are associated with **increasing** cholesterol in the blood, and heart disease.

Solid fats should be less than 7% of total food calories. Americans currently exceed the limit recommended for solid fats by **181%**

The risk for both cardiovascular disease and type 2 diabetes can be substantially **reduced** if saturated fats are replaced by mono- and polyunsaturated fats.

Sustainable grass-fed organic beef is **half as likely** to have 2 dangerous types of bacteria than grain-fed cattle.

Grass-fed cattle are leaner than grain-fed cattle and typically contain the level of omega-3 fatty acids. This is because the omega-3 fatty acid alpha-linolenic acid (ALA) is abundant in grass but not in grains.

Pigs reared outdoors until 80 kg of live weight or permanently have improved carcass characteristics with leaner meat and less back fat.

Restrictively fed outdoor pigs also have a higher proportion of omega-3 in their muscles.

Most people eat too much of the wrong type of fats, eating **20%** more saturated fat than recommended.

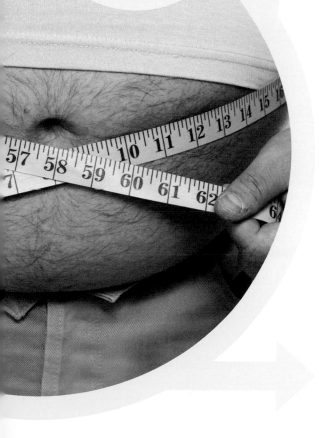

Salmonella is almost seven times more likely to infect large-scale and caged flocks of laying hens than smaller and non-caged flocks.

Average British supermarket chickens, intensively reared and fast-growing, contain proportionally **2.7 times** as much fat and 32% less protein compared to chickens in 1970.

'GROUND ZERO' FOR SWINE FLU
La Gloria, Mexico

When a young boy fell ill in the Mexican village of La Gloria in April 2009, few would have predicted that within a year, more than 18,000 people worldwide would be dead from the same virus.[13] But that's exactly what happened. Little Edgar Hernandez was 'patient zero' – the first confirmed case of swine flu.

La Gloria lies in the Perote Valley region of Mexico, eight kilometres from one of the biggest concentrations of pig farms in the world. It was the epicentre of one of the worst-ever health scares linked to factory farming. It spread much faster than anyone expected: within a week, ten countries were affected; it had spread to 180 countries by the end of August 2009.

In an area of just a few square kilometres in the valleys of Perote and Guadalupe, more than a million pigs are reared every year.[14] These animals spend their whole lives indoors in cramped concrete pens, their manure flushed into open-air lagoons.

When the pig farms first arrived, locals thought they would create jobs and boost the local economy. Guadalupe Gaspar, a villager who has campaigned for years against the expansion of the mega-piggeries, says that few local people work in them. "We were deceived,"

PATIENT ZERO A statue of five-year-old Edgar Hernandez stands in La Gloria as a reminder of the dangers of swine flu.

he says. "They said some companies were coming that were going to give people jobs. They never said anything about coming here to pollute."

Soon after the first pig farm arrived near La Gloria, villagers noticed a change in the quality of the groundwater. They complained to the local council and federal government, but their objections were ignored. Two years later, more than a quarter of the population of La Gloria – 28 per cent – went down with swine flu or something like it.[15]

Unusually, the viral strain had a mix of genetic material from two different swine flu viruses as well as from human and avian flu strains. Questions have been raised about the scientific link between the new virus in

La Gloria and the neighbouring piggeries. The company that owns the farms says health authorities carried out extensive tests on its farms and said there wasn't "any sign associated with the flu in our animals". Maybe it's just a coincidence that a new strain of flu, a key component of which was two different strains of swine flu, was first diagnosed right by some mega-piggeries? Either way, the focus of local protests about the farms shifted from pollution, smells and flies to disease.

FIGHTING BACK

Gaspar says: "When we realised what was happening, all the villages came together to demonstrate. People were dying. That is when we got up and started to fight; to make

INQUSITIVE PIGLETS
These intelligent, curious animals were born indoors and will spend their entire lives without setting foot outside.

CORPORATE FARMER
The Granjas Carroll de Mexico farming operation just outside La Gloria, Mexico, raises around one million pigs per year.

demands, so that the farms knew we didn't want them here. "We don't know when something else bad is going to happen to us. The government must get rid of the farms because while they remain, the pollution will continue and I am sure there will be more diseases," he warns.

Now the immediate panic is over, most people elsewhere have stopped worrying about catching swine flu. Yet there is every chance that the mix of conditions that created these strange new viruses will sooner or later throw up another pathological monster.

Dr Aysha Akhtar, a neurologist and public-health specialist and Fellow of the Oxford Centre for Animal Ethics, sums it up like this: "By confining billions of animals on factory farms, we have created a worldwide natural laboratory for the rapid development of a deadly and highly infectious virus."[16] Akhtar works for the Office of Counter-terrorism and Emerging Threats of the US Food and Drugs Administration. She points out that human terrorists don't have a monopoly on killing and causing chaos. Factory farming, she fears, has as much potential to do the job as any terrorist.

CHINESE FOOD SAFETY FEARS

The poisoning of Lake Taihu

China's fast-growing demand for meat to feed its middle classes makes it vital to any debate about the future of farming. But as demand for meat soars, so does concern about how food is produced there.

In recent years, there have been several major food health scares in China. In 2008, thousands of babies fell ill and six died when baby milk was contaminated with a chemical called melamine.[17] In 2011, people became ill after pig farmers decided to feed their animals Clenbuterol – an illegal body-building steroid that made the pigs grow huge without making their meat fatty. This illegal drug causes serious side effects, including palpitations and stiffening of the heart muscle.[18] It was another dangerous breach of public trust and echoed an incident five years earlier in which over 300 people were poisoned by meat contaminated with Clenbuterol.[19]

Today, it's factory farming that's putting public health at risk. And in China, factory farms come on a colossal scale.

Henan Province is the centre of China's pig-farming universe. One company, called Muyuan, rears more than a million pigs a year at 21 different sites. They are fed on, among other things, imported soya and fishmeal. Muyuan plans to boost its production to nine million pigs a year.

Kept indoors in giant sheds, the pigs' muck falls through slatted floors and is then slung into giant filthy lagoons, where it can seep

into groundwater and nearby crops. One local, who wants to remain anonymous, says his local Muyuan mega-piggery is ruining local roads and destroying people's water source. He says: "We never used to get mosquitoes, but now they're everywhere. We need nets just to open our windows. We used to sleep and eat outside when the weather was hot. That's impossible now – we have to shelter in our houses."

Lake Taihu, in Jiangsu Province, attracts swathes of tourists every year, drawn by its beauty and wildlife. But the lake – which supplies water to millions of residents in the nearby town of Wuxi – became so polluted

INDUSTRIAL FARM
A worker leaves at the end of a long shift at one of Muyuan's mega-piggeries.

that its water was undrinkable.[20] Even after being treated using the usual cleaning process, it was cloudy and smelled foul.

China's own figures show nearly ten Olympic-sized swimming pools' worth of nitrates from manure a year have been pouring into the lake. Today, it is dying.[21] A luminous green halo now haunts its shores. The stench can be overwhelming – a classic case of lethal algal bloom caused by nitrates in fertilisers and manure.

PIG IN MUCK
The muck from the millions of pigs often flows directly into local waterways.

ALGAL GROWTH
The water in Lake Taihu has been turned bright green by algae.

The table below shows the efficiency with which the calories and protein contained within human-edible grain is converted into animal products by intensive farming. In all cases, more than half is lost in the process.

Dairy

Eggs

Chicken

Pork

Beef

PLUNDERING PERU

The human cost of fishmeal production

Fishmeal is one of the filthiest secrets of factory farming. Millions of small fish are sucked out of the sea and crushed into fish oil and dry feed. This deprives millions of larger fish, birds and marine mammals of their food. It also pumps fatty waste into ocean bays, creating 'dead zones' where nothing lives.

Worst of all, there's a human cost – pollution from fishmeal factories has a devastating effect on the health of people living near them. These small, nutritious fish could be used to feed local people but instead they're shipped across the world to feed farm animals.

Peru is the world's second-biggest fishing nation. More than a billion pounds a year is made from the trade in anchovies, which are ground into fishmeal.[22] More than a million tonnes of the stuff is exported every year, making Peru the leading global fishmeal supplier.[23] A third of the 135,000 tonnes of fishmeal consumed in the UK in 2010[24] was imported from Peru.[25]

Chimbote, in Peru's Ancash region, is one of the nation's largest producers of fishmeal. Here, life expectancy is 20 per cent below the national average. Dr Wilber Torres Chacon, who works for the region's health department, explains: "The fishmeal production activity here causes several

> **The bay used to be considered the pearl of the South Pacific.**
> Maria Elena Foronda Farro

health problems: severe respiratory infections, asthma, acute diarrhoea, malnutrition, parasitic diseases …"

SKIN PROBLEMS

Dr Chacon says that as many as seven out of ten children in Chimbote suffer from skin conditions: this figure rises to 90 per cent for those living next to fishmeal factories. "These children are constantly exposed to fumes from the factories. You can see the blemishes on their skin," says Chacon. Infant malnutrition in the Ancash region is high – at between 20 and 30 per cent of the population. Health chiefs are trying to get locals to eat more fish, but this is a habit that seems to have died with the rise of the fishmeal.[26]

SICK CHILDREN
Dr Torres Chacon examines a child in Chimbote. Skin conditions are widespread.

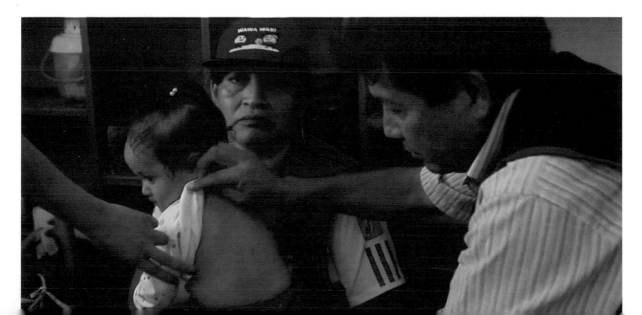

FACTORY FUMES
A fishmeal factory in Chimbote belches toxic fumes into the atmosphere.

Some locals have found their own way of making a modest living from the fishmeal factories. Around a thousand people are thought to earn money by siphoning fat from fishmeal factory sewage pipes and selling it. Although illegal, this practice is actively supported by the fishmeal companies.

The foreman of one of these siphoning teams says: "It is not possible to clean up the bay. This is all we can do. We need help to get out of here. Our children are suffering; they are being poisoned. The only reason we don't leave is because we can't afford to." Local campaigners have worked tirelessly to stop

World fish catch

1/5

of marine stocks
are fully exploited.
28% are overexploited, 3% are depleted
and 1% are recovering from depletion.

Peru exports
half its fishmeal.

of its infants are
malnourished.

pollution from fishmeal. Maria Elena Foronda Farro has fought against corruption and environmental damage linked with the industry since the 1990s.

This award-winning campaigner says: "Our first victory was forcing a number of the fishmeal factories to relocate out of the centre of Chimbote. We managed to get an order against them in 2009, and in 2010 they began to move. It was a big thing – they had to build entirely new plants."

According to Foronda, 26 companies already located in the industrial area

were given until the end of 2010 to clean up their technology. As of March 2011, only eight had done so.

"Just as we thought we were getting somewhere, the regional government started authorising the building of new low-grade fish-processing plants, arguing that it would create jobs," she explains. "The fishmeal business has left us without much to live on, plundering our natural resources and failing to put anything back into development in Chimbote. The bay used to be considered the pearl of the South Pacific. Look at it now."

To produce
1 tonne of farmed
salmon requires

of wild fish. 1 tonne of farmed
trout requires 2.5 tonnes
of wild fish.

Fish farms across the world consume a large proportion of the total fish catch. These fish could have been fed to people or left in the ocean to feed wild fish and other sea creatures.

"Our children are being poisoned. The only reason we don't leave is because we can't afford to.

Resident of Chimbote, Peru

PERUVIAN COASTLINE
A dolphin pod swim along the coast of Peru. An intensive chicken farm sits on the barren landscape behind them.

EXPANDING WAISTLINES
The health costs of poor-quality meat

If current trends continue, half of all Americans will be obese by 2030. The predictions for the UK are almost as bad. Experts think this soaring obesity epidemic will cost an extra $48 billion in the US, or nearly £1.25 billion in the UK.

Factory-farmed food has played a key role in the global obesity crisis. Scoffing roast chicken and sausages might seem healthier than downing cake, but evidence shows factory farming has stripped away most of the goodness of cheap meat. At the same time, the fat content has soared. Intensive farming has had such a big impact on meat quality that some say you'd have to eat four whole factory-farmed chickens to get the same nutrients you would have got from a single chicken in the 1970s. A portion of today's supermarket chicken contains 50 per cent more calories than it did in 1970.[28] Professor Michael Crawford, of London's

Institute of Brain Chemistry and Nutrition, says: "The intensification of animal farming has virtually destroyed the nutritional quality of our food." He describes modern industrial chicken farming as "fat production, not meat production". His study, published in *The Lancet* more than half a century ago, showed that the ratio of 'bad' to 'good' fats in farmed animals was 50:1; compared with less than 3:1 in their wild counterparts.

Now, the situation is much worse. Factory-farmed animals are more or less 'selected for obesity' and get virtually no exercise so, as Crawford puts it: "If you eat obesity, you become obese." Most people eat far too much saturated fat, found in fatty cuts of meat as well as sausages and pies. This overload is linked to high cholesterol and heart disease. Ideally, people should eat about the same amount of omega-3 ('good') fat as omega-6 (saturated fat).[29] Current advice allows for a bit of extra omega-6 – at four times more than omega-3. Yet most Western diets include as much as 10–25 times more omega-6 than omega-3.[30]

Strong evidence links this to the massive shift from natural grazing on farmland to rearing animals on grain.[31] Unlike grain, grass is full of omega-3. Studies show that when beef cattle are fed on grass, their meat contains more omega-3 than those fed on higher levels of grain or soya, as happens on factory farms.[32]

Evidence shows that keeping animals in higher-welfare conditions provides more nutritious food. A major review of scientific data found that eggs, milk and meat from animals kept in better conditions often contained less fat and more nutrients than factory-farmed food.[33] Farmed chickens used to be active and eat vegetation and seeds. Today's factory-farmed chickens, by contrast, can barely move, yet are fed high-energy foods.

"Such chickens are no longer a protein-rich food, but a fat-rich food," says Crawford. "The explanation is simple, namely that they are fed largely on cereals." The 2004 film *Supersize Me* followed what happened when filmmaker Morgan Spurlock spent 30 days eating nothing but food from McDonald's. He piled on an incredible 11 kg (24.5 pounds);

FAT CHICKENS
Today's factory-farmed broilers are bred to be fast-growing and produce meat that is high in fat and low in protein.

The true cost of the Big Mac

$4.56 RETAIL COST

$0.38 in cruelty. A total of $20.7 billion in cruelty costs is imposed on Americans each year.

$0.67 in environmental losses. A bill of $37.2 billion in environmental costs is related to US animal food production each year.

$0.70 in subsidies. US taxpayers pay $38.4 billion in government subsidies to fund the meat and dairy industries each year.

$5.69 A chunk of the $314 billion in healthcare costs incurred annually to treat conditions related to meat and dairy consumption.[34]

TOTAL COST: $12.00

his body mass index rose by 13 per cent; his cholesterol was up and he suffered mood swings, sexual dysfunction and a build-up of fat in his liver. It took Spurlock 14 months to lose the weight.

CHANGING DIETS

In rich countries, most people eat about 200-300 grams of meat per day. Experts say this needs to drop to around 90 grams per day for both public health and environmental reasons. The benefits, they argue, would include a drop in the risk of colorectal cancer, breast cancer and heart disease.[35]

Farm animals never used to compete with people for food. Cows would eat grass and pigs and poultry would eat scraps and leftovers and forage. Now, more calories go

GRASS-EATERS
Cows that feed on grass produce meat containing healthier fatty acids than their grain-fed counterparts.

into factory-farmed animals than come out. We convert food that people could eat, like grain and soya, into meat through factory farms.

There are signs that meat-eating in the West may have already peaked. Meat consumption in the US has declined by 12 per cent in recent years.[36] More and more people seem to be taking up the idea of eating less meat. UK initiatives like 'Meatless Monday" are growing in popularity[37] and Americans are reportedly warming to the idea of 'flexitarianism' – simply eating less meat.

MUCK
Killing the world's waterways

When you separate animals from the Earth and you get too much manure in one place, rivers and lakes often get polluted. All this muck presents the single biggest threat to our waterways. Fertilisers and manure spread on the land can be washed into rivers. Manure contains massive amounts of nutrients, which suck oxygen out of the water. It literally chokes the life out of aquatic animals and plants, creating dead zones where nothing lives.

Sometimes, this causes deadly problems for the people who live near intensive farms. From the beaches of Brittany in France to the rivers of North Carolina, we find people suffering because of mountains of muck.

Brittany, for example, produces 14 million pigs a year.[38] More than half of the nation's pigs are reared there,[39] yet you'd be hard-pressed to spot a single one rooting

↑ POISONOUS LAKES
Toxic pig sewage collected in lagoons often leaks out to pollute waterways.

around in a field or farmyard. One summer recently, the corpses of 36 wild boars, a badger and a river rat were washed up on the Saint Maurice beach near the mouth of the Gouessant River within a few days of each other. All but one had lethal doses of hydrogen sulphide in their tissues. Authorities confirmed they were killed by toxic algae,[40] which flourishes when too much nitrogen is carried downstream from polluted rivers and waterways. As the algae dries up it releases hydrogen sulphide, known as 'sewer gas', and other toxic fumes. "Nothing lives in the bay of St Bricuc anymore," says Yves-Marie Le Lay, who runs a local conservation group. "Everything has died."

TOXIC PLANTS

When the contaminated weeds wash up on land, they can also be deadly for humans. In 2009, a man involved in cleaning up

Brittany's toxic seaweed was found dead by the foot of his truck. It turned out Thierry Morfoisse had died of cardiac arrest, possibly resulting from a pulmonary oedema – the often-fatal sign of hydrogen sulphide poisoning. Le Lay says massive cereal plantations have replaced the rolling pastures that animals once grazed on in the region. Now, the animals live in indoor 'feedlots', and the cereals are turned into animal feed, which is supplemented by foods imported from as far away as Brazil. Across the Atlantic, the same problem with muck exists. In one North Carolina county alone, 2.2 million pigs generate as much untreated manure as the sewage from central New York City.[41] Like the muck produced on every factory farm, it has to go somewhere – and often ends up in the wrong place.

Huge lagoons of pig sewage, more than 3,000 of them,[42] dot the North Carolina landscape. Each one holds masses of

'Nothing lives in the bay of St Brieuc any more. We used to fish periwinkles and clams from the rocks. '

Yves-Marie Le Lay

excrement, urine and other things like blood, afterbirths and stillborn piglets. The liquid is pinkish brown. Anyone who falls in is most certainly doomed. *Rolling Stone* magazine reported some years ago that five members of the same farming family were killed after one was overcome by fumes and toppled into one of these lagoons. The others were all killed as they tried to save him.[43]

DYING FISH

Sometimes the lagoons overflow, spilling the toxic soup into fields, where it seeps into groundwater. To avoid that, some is pumped out over nearby farmland, but when it's not done carefully the land becomes covered in stagnant pools. Scientists believe this is what caused the North Carolina's River Neuse to become polluted.

Former marine Rick Dove once earned his living from the Neuse. He and his son ran a thriving fishing business together as well as their own seafood shop. But just a few years later they were forced to quit after fish began to die and they themselves fell seriously ill. They were under threat from a deadly organism – *Pfiesteria piscicida*. Between 1991 and 1997, this single-cell organism is thought to have killed more than a billion fish in North Carolina. It also made

fishermen sick.[44] Its arrival in the Neuse was traced back to pig manure. "Between 1991 and 1999, we lost billions of fish – certainly a billion and a half. I know that because I was on the river. I watched," says Dove. "These fish had huge holes right through their body. They'd get sores, these bleeding, ulcerating sores, and fishermen get the same sores."

Once he recovered, Dove went on to become a river keeper on the Neuse and spent many years investigating who was responsible for the pollution. After years of legal battles, there are new regulations in North Carolina governing the disposal of pig manure, but

MUCK FACTORIES
The muck from factory-farmed pigs collects in huge quantities in a small area.

this environmental nightmare is not confined to a single state. Where there are factory farms, there are all too often problems with getting rid of the muck. The result is widespread toxic pollution of rivers and lakes.

It's no surprise that a proposal for Britain's first US-style mega-dairy was dealt a death blow by the Environment Agency's concerns about water pollution. Near and far, the effects are being felt.

CHAPTER 3
WHY ANIM
MATTER

ALS

INTRODUCTION

Worldwide, about 70 billion farm animals are reared for food each year. More than 50 billion of them live in factory farms. Held in cramped, crowded conditions and pushed beyond their natural limits, they are often treated like nothing more than machines before being slaughtered

Yet these are sentient creatures that feel pain and fear.

Factory farming is inherently cruel. Whether it's egg-laying hens in cages so small they can't even flap their wings, broiler chickens collapsing under the

↓ PIGS SUCKLING
Piglets form a hierarchy early on as they suckle from their mother. Pigs are sociable animals that need company.

weight of their supersized bodies or cattle with no access to grass on a gigantic Argentinian feedlot, intensive farms cause animal suffering.

Overcrowded, barren, indoor systems simply don't provide a good enough environment to keep animals healthy and happy. All over the world – from California's mega-dairies to Taiwanese farms – billions of animals are suffering every day because of factory farming. But why does it matter, and why should we care how animals are reared, transported and slaughtered?

WHAT IS ANIMAL WELFARE?

The five freedoms

Animal welfare means different things to different people. Welfare generally means the 'quality of an animal's life, as it is experienced by an individual animal'.[1] It includes psychological wellbeing and the ability to express natural behaviour. Welfare can be described as high if the animals are fit and healthy, feeling good and free from suffering.[2]

The five freedoms are an aspirational but useful guide for anyone responsible for looking after animals. First developed in the UK during the 1960s, the five freedoms are now used as a framework by animal welfare organisations, governments and individuals around the world to explain how good animal welfare can be achieved.

The five freedoms are:

1. Freedom from hunger and thirst – by ready access to free water and a diet to maintain full health and vigour.
2. Freedom from discomfort – by appropriate environment, including shelter and comfortable resting area.
3. Freedom from pain, injury or disease – by prevention and rapid diagnosis and treatment.
4. Freedom to express normal behaviour – by providing sufficient space, proper facilities and company of the animal's own kind.
5. Freedom from fear and distress – by ensuring conditions and care that avoid mental suffering.[3]

Meeting these freedoms is essential in order to avoid suffering and poor welfare. But there is increasing recognition and awareness that animals also need to experience positive emotions to have good welfare. Farm animals are sentient beings, which means they have feelings that matter to them. It is a concept recognised by EU law and means that any new legislation that affects animals must take this into account.

People's views on animal welfare tend to be influenced by deep-rooted cultural beliefs. A person with an 'industrial' view values life improved through science and technology and believes animal welfare is achieved through health, biological functioning, productivity and control over nature.

In stark contrast, a person with an 'agrarian' view values a traditional life, in harmony with nature, and believes animal welfare is

DUCKS IN WATER
Ducks are aquatic birds that need access to water in order to preen and stay warm.

achieved through attention to the emotions and freedom of the individual animal.

There is, however, a compromise – a 'professional' view. This is about using skills and science to improve practice and setting publicly acceptable standards. If producers move away from the agrarian view and adopt the professional rather than the industrial approach, it will improve both animal welfare and public trust as well as being humane and sustainable.

PREGNANT SOW
Given space to roam outdoors, sows build themselves nests in which to give birth.

WHY DOES ANIMAL WELFARE MATTER?
Changing values

The idea that society should act to protect animals was first enshrined in law as far back as 1822. The world's first animal welfare legislation was an act of parliament to protect UK farm animals. Richard Martin MP's Act to Prevent the Cruel and Improper Treatment of Cattle stated that if any person or persons "shall wantonly and cruelly beat, abuse, or ill-treat any horse, mare, gelding, mule, ass, ox, cow, heifer, steer, sheep, or other cattle", they would be fined. Failure to pay the fine would result in a prison sentence of up to three months.

Laws to protect UK pets and outlaw cruel sports followed in 1835, and since then many acts have been introduced to protect wildlife, ban specific acts of cruelty and update existing legislation. Many governments around the world have brought in their own laws to prevent animal cruelty – some going further than others.

These laws were passed because people cared enough about animals to make it happen. Attitudes towards animals are often influenced by education and social and cultural factors, but few people would argue against the need to protect animals from unnecessary suffering. Most people would agree that animal cruelty is morally wrong.

Throughout the world, many millions of people support animal protection organisations – some give money to fund activities while others are motivated enough to give hands-on help or take part in campaigns and events to highlight animal welfare issues.

A survey by the European Union found that 94 per cent of people living in the EU think protecting farm animal welfare is important with 82 per cent believing that farm animals should be better protected than they are

now. The majority are prepared to pay more for products sourced from higher welfare production systems and look for animal-friendly labels.[4]

In the UK, public outcries over farm animal suffering, led by celebrity chefs Hugh Fearnley-Whittingstall and Jamie Oliver, have led to many more people buying animal-friendly products. While the lives of countless animals have without doubt been transformed for the better by modern legislation, it's 'modernisation' itself that has caused misery for billions of farm animals through the industrialisation of farming.

UPDATING THE LAW

As scientific knowledge of animals improves, laws should be routinely updated to reflect that new awareness. For example, in 1997, the EU agreed to recognise the scientific

'Humanity's true moral test consists of its attitude towards those who are at its mercy: animals.'
Milan Kundera

evidence that all animals – from pets to farm animals – are sentient beings that can experience emotions like pain, fear and joy. This legally binding protocol attached to the 1997 Treaty of Amsterdam was introduced in response to compelling scientific evidence. It was further strengthened in 2009, and now the welfare of animals must be taken into account when any new EU legislation is passed.

When faced with scientific research showing, for example, that pigs can identify faces to distinguish different people,[5] cattle value social contact with others and can remember 50–70 individuals,[6] and chickens can solve mazes to be allowed access to dust-bathing material or a nest,[7] the EU was forced to act. Animals have evolved to cope with life in the wild. Thousands of years of domestication of farm animals have done little to change their basic motivations and behaviour patterns.

Intensive farming fails to appreciate animals' needs and their capacity to suffer. This means that billions of sentient animals are routinely subjected to pain and deprivation in a variety of systems, from mega-piggeries in China's Henan Province to broiler chicken farms in America's Deep South.

NATURAL ROOTERS
Boars evolved in the wild to eat a wide variety of food found while rooting and foraging. Their relatives, domestic pigs, have preserved the instinct to root.

SOUTHERN DISCOMFORT
The rise of the industrial chicken

There are more chickens in the world than any other farm animal. Worldwide, 55 billion chickens are reared for meat each year; nearly three quarters are 'factory farmed'. The vast majority of these birds will never see daylight during their short lives. Each year some 27,000 broiler

A typical farm raises 600,000 birds a year, and with that number of chickens comes an awful lot of muck. The chickens of Maryland and Delaware alone generate 42 million cubic feet of litter, enough to fill the US Capitol dome nearly 50 times over.

Georgia is the largest producer of meat chickens in the United States, rearing 1.4 billion every year.[10] If it were a country, it would be the sixth-largest poultry producer in the world.[11] The heart of the industry is in the foothills of the Blue Ridge mountains in a place called Gainsville. It enjoys the dubious title of 'poultry capital of the world'.

Georgia's chicken industry started during the Great Depression when a struggling Gainsville feed salesman, Jesse Jewell, started selling baby chicks to farmers on credit. The farmers raised the chicks and then sold the fully grown birds back to him for a profit.

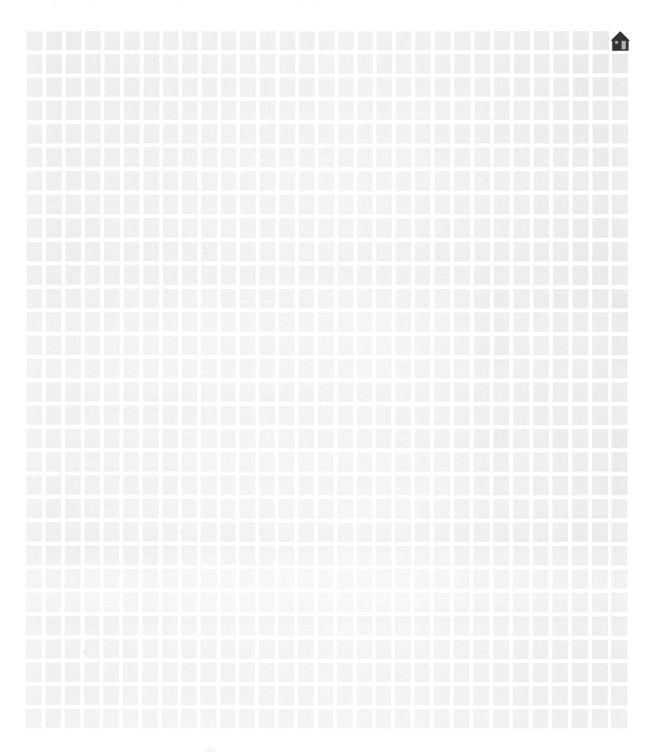

GHOST ACRES
90 hectares (220 acres) of arable land are needed to produce the
cereal to feed the chickens in one 892 m² (9,600 sq ft) shed.

Seven-week lives

Factory-farmed chickens reach their
target weight in no more than seven weeks.
At this age, they are still very young;
chickens don't reach puberty and start laying
eggs, for example, until about 18 weeks old.

A sheet of A4
Factory-farmed chickens are kept in extremely crowded conditions.
Each bird has what has become the customary floor space,
the size of a sheet of standard A4 paper.

INTENSIVE CHICKEN FARM
One 892 m² chicken shed produces 150,000 chickens per annum.

CONTRACT FARMS

Eventually, Jewell had enough farmers rearing broilers for him that he was able to open his own processing plant and hatchery.[12] He was a pioneer of so-called 'contract farming'. By 1954 he had added a feed mill and rendering plant. Eventually, J. D. Jewell Incorporated had the resources to manage every phase of chicken production, from hatching to processing, distribution and marketing.

For the next couple of decades, broiler production in Hall County continued to thrive. Sensing a quick buck, everyone piled in. Competition drove prices down, but during the 1970s and 1980s American demand for chicken rose. In the 1990s, the number of companies involved fell – there were fewer but bigger operators, serviced by an army of contract growers. Today, just a few companies dominate the industry.

Carole Morison, an award-winning broiler farmer whose contract was terminated after 23 years, has told how she was constantly bullied by the companies she worked for. This is the way 99 per cent of broiler chickens in America are reared – by producers at the beck and call of company bosses in far-off offices.

For the chickens themselves, it's a grim existence, with virtually no legal protection. All farm animals are exempt from America's Federal Animal Welfare Act. Despite making up 95 per cent of farm animals in the US, chickens aren't even protected by the Humane Methods of Slaughter Act.

No other farm animal has been as selectively bred as the meat chicken to reach an unnatural size so quickly. Over the last 50 years, growth rates have quadrupled.[13] Getting the birds to target weight now takes

no more than seven weeks.[14] This rapid growth has allowed for mass production of cheap meat, but the chickens pay a heavy price. They suffer leg problems, heart disease, lung problems and a condition dubbed 'flip-over syndrome' where chickens suddenly start frantically flapping their wings, lose balance and within about a minute collapse and die.

Like other factory-farmed animals, chickens are kept in extremely crowded conditions. The typical 'grow house' for a Georgian flock is reported to be 15 metres wide and 150 metres long[15] and holds more than 30,000 chickens. Each bird has a space the size of an A4 sheet of paper. They are kept in continuous light because it makes them eat more and therefore grow faster. In fast-growing breeds, the development

of the birds' bodies can't keep pace with their weight gain, making walking painful. Most only move around when absolutely necessary to reach food or water. Towards the end of their short lives, they are mostly forced to sit or lie down. Some are in such poor condition they can barely walk.

An estimated 42 million chickens die in Georgia every year before they reach slaughter weight. The ones that make it to slaughter are manually 'grabbed' and taken to the abattoir. 'Catchers' grab up to seven chickens at a time – three in one hand and four in the other – and push them into crates for loading onto trucks. According to the Southern Poverty Law Center, a civil rights organisation based in Alabama, catchers are expected to grab and crate birds at a rate of around 1,000 an hour.[16]

GROWING CHICKS
The farmers receive the chicks from a hatchery at one day old. In their first week of life they will increase in size by 300 per cent.

develop a condition known as 'claw hand' from gripping so many chickens so tightly over many years. As few as five per cent of catchers are thought to have clear employment conditions and many travel from job to job in unsafe, unlicensed vehicles. Fewer than 15 per cent of crews keep proper records to ensure that workers are paid properly for the hours they do.[18]

One of the worst jobs is live hanging. This is where each bird is lifted by its legs and hung on hooks, at shoulder or head level. Around one in seven workers is injured on the job, more than double the average for all private industries.

At the slaughterhouse, the birds are typically loaded onto conveyor belts and shackled upside down by their legs. They may be stunned – rendered unconscious – before having their throats cut, but it's not required by US law. On welfare grounds alone, there is a clear case for changing the way meat chickens are reared, but it isn't only the birds that suffer. According to a report by the US Bureau of Labor Statistics, poultry processing is one of the industries with the highest rates of non-fatal illnesses among workers.[17]

Georgia offers a depressing insight into the working lives of the 47,000 people employed in the industry, a high proportion of them female and Latino. Being a catcher is particularly tough. Their hands are often swollen to double the normal size, and some

Factory-farmed chickens are not unique to the US. Four out of five UK chickens are reared in a similar way and it's much the same throughout the rest of Europe and the rest of the world. Recent investigations exposed the awful conditions chickens are forced to endure in the UK and the outcry that followed, spearheaded by celebrity chefs, has led to record numbers of people demanding higher-welfare chicken.

Now, nearly a quarter of chicken sales are from farms keeping birds in better conditions – whether it's free-range, organic or RSPCA Assured. There are also signs of positive change in Georgia, where higher-welfare meat labelled 'Raised with care' or 'Great tasting meat from healthy chickens' can now be found in supermarkets.

CHICKENS – WHERE THE EGG COMES FIRST
Caged egg-layers

Although battery cages have been banned in the EU, around 60 per cent of the world's 6.5 billion laying hens live in cages. Around 30 million of them are in Taiwan, where laying hens generally live in one of two types of system – one known as 'traditional' and the other known as a 'controlled environment'. In both types, the hens are caged.

On 'traditional' farms, hens are usually kept in row upon row of cages, covered by a tin roof but otherwise open to the air. The wire cages are barely bigger than the hens crammed inside and are completely bare. The hens have nothing, except access to a trough and water, the absolute basics of life.

Conditions on 'controlled environment' farms are little different, except that the barns are completely enclosed, with temperature and ventilation controlled by computers and fans. These farms exist all over the world, and their supporters claim they are somehow safer and healthier because the birds aren't exposed to agents of sickness and disease outside. Yet this myth was exposed in Britain when an intensive Bernard Matthews turkey plant was hit with highly pathogenic avian influenza. It showed that being 'closed' to the world and run by computer does not make a shed immune from the obvious laws of nature.

Taiwan also has some open-air farms. The birds are still crammed into the most appalling tiny cages, but at least they get natural light and can feel a cooling breeze in what can often be intense heat.

At the country's only organic farm, it's a world away from what you'd see at a European organic farm. Instead of scratching around in the open air, some 300,000 chickens are kept in cages stacked seven high, in four industrial buildings. The word 'organic' is simply related to the feed they are given.

EGG MACHINES
Laying hens stand in rows of cages, with only the bare necessities for life.

In overcrowded battery cage systems like these, birds often get a space no bigger than a hand to live in. They are unable to flap their wings or even turn around. Nor can they perform natural behaviours such as nesting, dust bathing and foraging. Instead, they often turn on each other and peck at each other's feathers, sometimes even resorting to cannibalism. Factory farming's answer to this is to chop off the end of the birds' beaks with red-hot guillotines or infra-red lasers when a few days old. Since beak tips have nerve endings, this practice causes both immediate and lasting pain.

In systems like this, it's not just eggs the staff have to collect. Birds often die and their bodies are simply pulled out and chucked into bins. Yet it's not just dead birds that suffer this fate. At one Taiwanese farm, birds that didn't produce enough eggs or that were very sick were seen stuffed into plastic bags until they suffocated, then thrown in the bin.

FORCED MOULT

In Europe, laying hens are usually slaughtered after one year. This is because they renew their feathers every year and stop laying eggs for a few weeks while that happens. In Taiwan, hens are typically kept for two years and to reduce the drop in production during feather renewal, they 'force-moult' the birds. This involves starving them for ten days, which speeds up the process and gets them laying eggs again more quickly. After that, they're caged for

another year before they're slaughtered. While banned in Europe, 'forced moulting' is still allowed in the United States and in many other countries.

The question is not "Can they reason?" nor "Can they talk?" but "Can they suffer?"
Jeremy Bentham

110

Conditions in Taiwan's egg farms are not unique. Life is very similar for almost 4 billion caged laying hens worldwide. Hens need access to the open air, daylight and space to exercise and move around freely. They are naturally inquisitive animals and it's important they are also able to express their natural behaviours of perching, dust bathing and foraging. While standard battery cages have been banned in the EU since 2012, the law still allows for so-called 'enriched' cages which gives each bird just a

↑ DUST BATH
Dust bathing is an important part of a chicken's natural behaviour.

postcard-sized area of extra space. They get a place to perch and scratch, but they still never see daylight and have to stand on or above a sloping wire floor to make egg collection easier. So while the EU ban was a big milestone, the reality is there's still a long way to go in Europe and elsewhere before hens have truly better lives.

GAUCHOS ON HORSEBACK
The traditional image of Argentinian beef farming belies a harsh reality.

BATTERY BEEF
Feedlot farming

When most people think of Argentinian beef, they think high quality, high welfare. The reality is that while many Argentinian cattle do graze on lush grass, a growing number are now reared in vast, thousand-strong feedlots, where they don't have access to a single blade of grass.

'Four thousand cows on one field is too many.'
Miguel Martinez

Just as has happened elsewhere, farm animals in Argentina are being taken off pasture and put in intensive systems where their food has to be provided. It's the worldwide demand for soya used in livestock feed that is shaping the Argentinian landscape today. An estimated 200,000 hectares of woodland are believed to be lost each year to make way for soya.[19] GM soya now covers at least 19 million hectares (47 million acres) of the country[20] – 65 per cent of the entire farmland[21] – much of it for export. In many parts of the countryside, field after endless field of soya plants can be seen.

Argentina now offers a double window on the world of industrial farming. On the one hand, it's literally feeding factory farms in the UK and other parts of Europe via shiploads of ground soya. On the other, it's a showcase for one of the most intensive farming systems in the world: feedlot farming, or what some call 'battery beef'. This is farming on a massive scale. Thousands of cows are often kept crowded together in huge pens and are forced to stand on thick brown mud and excrement. Some can be seen walking through mud and excrement up to their stomachs.

NO GRAZING

Cattle will naturally graze for up to nine hours a day. On feedlots there is no grass to be seen. This means the cattle not only get bored but are unable to express their natural grazing behaviour. On many feedlots, there is also no shade or shelter for cattle, sometimes exposing them to blistering heat and sun. Stocking densities are also a problem. To ensure each animal has enough space, there should never be more than two cows per acre, but on feedlots it's many more per acre. Cattle form long-lasting social groups, and problems like aggression can arise if different groups are forced to mix. In feedlots, hundreds of cattle are usually randomly put together in huge muddy pens.

But it's not just the cattle that are suffering. People who live near these feedlots have reported an overwhelming stench engulfing their homes as well as problems with rats and flies. Miguel Martinez, a civil servant, grew up in the area and planned to build a retirement home there on land he inherited from his grandparents. That was until he realised the land next door to his had been turned into a feedlot. Where once there had been a duck

# of all Argentinian farmland is soya fields

pond, woods and a fig tree, now stood thousands of cattle in nothing but mud and muck. "I'm not against progress or anti-development," says Martinez. "I don't mind living among cattle and horses. But four thousand cows on one field is too many, and the consequences are far-reaching. Small farmers cannot compete; the water supply is polluted; the air is bad."

Already rife in South America and the US, these cattle feedlots could soon make their way to the UK and other parts of Europe.

Meanwhile, in restaurants across the EU, diners continue to tuck into bargain-basement beef from Argentina thinking it's some kind of luxury product. Most are blissfully ignorant of the conditions in which it was produced and know little about the true quality of the meat.

In the UK, most cattle are still fattened on grass but across Europe many herds are brought indoors, where they are often forced to stand on uncomfortable slatted floors.

CAGED PIGS
China's growing pork industry

Half the world's pigs live in China. In recent years, the Chinese have invested massively in intensive farming systems developed in Europe and the US. The British government signed a multi-million-pound deal with China in 2007 to export thousands of live breeding pigs to the country.[22] Chinese farmers have been chartering Boeing 747 planes to fly the live animals 9,000 kilometres (5,500 miles) across the globe.

Whether it is steamed, roasted, barbecued or minced in dim sum dishes, the Chinese are big on pork, eating 34 kg (75 lb) per person per year. In Britain, it's just 25 kg (55 lb) per head. Although many pigs are still reared on traditional smallholdings, big players in China's food-production industry are eagerly importing the most intensive techniques they

can find in the West. They see the UK and US as role models. Due to recent food scandals, the Chinese now seem to be more open to animal welfare ideas – at least, officially.

Muyuan is one of the biggest pig producers in China. It has no fewer than 21 pig farms in Henan Province – the centre of China's

pig farming universe. It has been supported by the International Finance Corporation, the private lending arm of the World Bank.[23] This means that, indirectly, it is subsidised by taxpayers around the world.

By 2017, Muyuan expects to be rearing up to nine million pigs a year, in five different regions – more than the entire British pig industry. Some of its breeding pigs are imported – most recently from Canada – because of the more commercially productive pig breeds available abroad. They are fed on imported soya, fishmeal from Peru, Chinese wheat and added vitamins and minerals.

Sow stalls are routinely used. These are narrow, indoor stalls with metal bars, in which pregnant sows are held until they give birth. They stop pigs being able to move around or go outside and they can cause lameness due to muscle wastage. Research

↑ SOW STALLS
Pregnant sows are kept in narrow stalls that prevent them from turning round.

shows that these sows show signs of frustration like biting the bars of the stalls. Just before giving birth, the sows are moved to farrowing crates, which are similar to sow stalls but with a space for the piglets. Bars, to stop the piglets being crushed, keep the sow out of the piglets' lying area. Farrowing crates prevent a sow from expressing her natural nest-building behaviour and stop her being able to move away from the piglets, even when they bite her teats. Some farms just trim the piglets' teeth without anaesthetic.

In factory farms, growing pigs are kept indoors all the time, often on bare concrete or slatted floors with no bedding. It means they can't express natural behaviour like foraging. They become so bored that many fight each other. Most have their tails cut off to stop tail biting, but it's painful and can cause long-term suffering. The systems are so automated that just one member of staff can 'take care' of 3,000 pigs. Once they're ready for slaughter, most of the Muyuan pigs are loaded onto lorries and sent to major towns in China. This means they are loaded onto open-sided lorries and have to endure journeys of up to 30 hours.

HEALTH CONCERNS

More and more of these mega-farms are expected to appear in China to meet the growing demand for meat, unless Chinese consumers start flexing some economic muscle. The food scares have left them understandably nervous, but if they start choosing healthy foods, the industry will be forced to respond. An unhappy pig is an unhealthy pig, and an unhealthy pig makes unhealthy food.

Sow stalls are now banned in Sweden and the UK, and their prolonged use is prohibited throughout the rest of the EU. They are being phased out in certain US states as well as in New Zealand, and voluntarily phased out in Australia. Many food companies are also rejecting them thanks to consumer pressure. Farrowing crates are still routinely used throughout the world, except on higher-welfare farms.

BARREN LIVES
These piglets will spend their entire lives indoors, unable to root or forage.

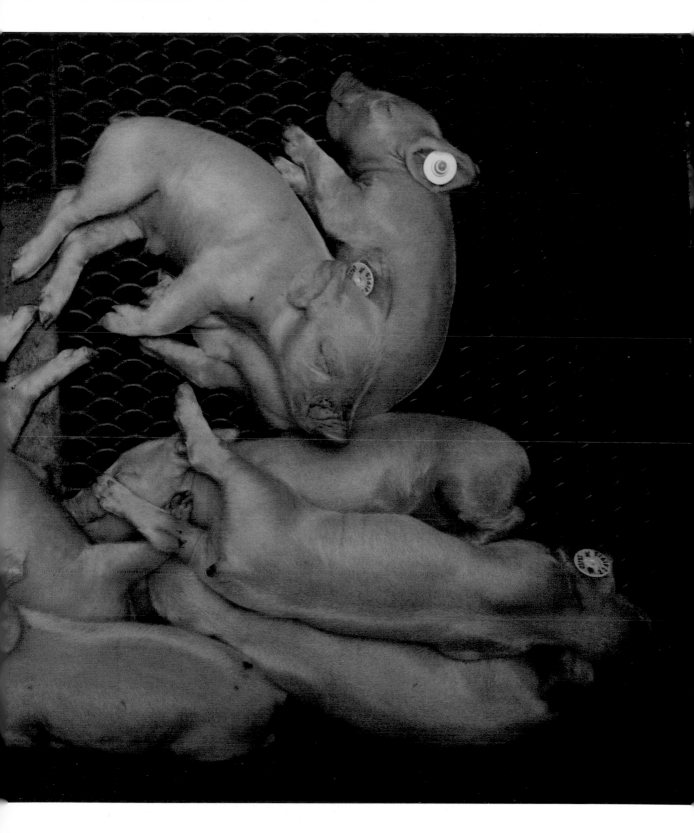

CHAPTER 4
RESOURC

ES

INTRODUCTION

The trade in cereals and soya to fuel factory-farmed animals is booming. But producing and transporting all this animal feed is a dirty, thirsty business that's fast using up precious natural resources like land, oil and water. People are being pushed out to make way for massive cereal and soya farms run by multinationals. The very poorest communities are under threat.

Over the next few decades, the world's livestock population is set to nearly double as the global demand for meat soars. Even more land will be needed to grow feed for these animals. It comes at a price.

LOST TREES
In Argentina, 200,000 hectares (500,000 acres) of woodland are lost each year to make way for soya.

'When the multinational arrived to grow soya, they fenced off the land and installed armed guards. What could we do?'

indigenous tribesman

LAND GRAB
Factory farming's ghost acres

Factory farming uses more land than meets the eye.[1] As agriculture became more intensive, the amount of land used to grow crops also went up, partly because factory-farmed animals are fuelled by the mass production of feed crops. If all the crops grown to feed factory-farmed animals were put into one field, it would cover the whole of the European Union or half the United States.[2]

A third of today's global cereal harvest goes to livestock; even more – 70 per cent – in rich countries.[3] Ninety per cent of the world's soya meal is destined for industrial livestock. Every year, an area of forest half the size of the UK is torn down, mostly to grow animal feed and for cattle ranching. So factory farming uses 'ghost acres' – it's just that the animals are often no longer on the land.

The term 'ghost acres' is used to describe the difference between the amount of food a country needs and the amount it grows on its own land. This gap is usually filled by imports. Factory farming has made the amount of 'ghost acres' dramatically shoot up. In the last 40 years, the amount of land used worldwide for agriculture has increased by almost 500 million hectares (1.2 billion acres) – ten times the size of France.[4] Land is being swallowed up at an alarming rate and most of the fertile land that's left is in South America and Africa, where wildlife and local people will likely be thrown out to make way.

Take Argentina, where around 200,000 hectares (500,000 acres) of woodland are lost each year to make way for soy.[5] Genetically modified soya now covers at least 19 million hectares[6] of the country – 65 per cent of the entire farmland – much of it for export.[7]

DISPLACED PEOPLE

For centuries, the American Indian Toba Qom tribe lived deep in the 'impenetrable' forest of northeastern Argentina. Today, only a few remain as most of the ancient woodland has been felled for timber and their land carved up to grow soy. "Over the centuries, our people have been pushed into smaller and smaller territories", says Abel Paredes, secretary of the American Indian Toba Qom community. "Then some years ago a multinational came in and bought our land. The provincial government sold our land, with us included in the price, because we happened to be there. We used to live in the forest, hunting for alligators, honey, iguanas, fish, and we tilled the land, growing vegetables and cotton. When the multinational arrived to grow soya, they fenced off the land and installed armed guards. What could we do?"

Without land for hunting and growing fruit and vegetables, many of the Qom have been forced to move to cities, where conditions are tough. For those few left, life is a bitter struggle. Some have even starved to death.[8] The treatment of the Qom people is just one example of how the poorest and most vulnerable are being cast aside to ensure a steady supply of artificially cheap, poor-quality meat for people thousands of miles away.

⬆ **NO PLACE TO GO**
The Qom have been forced off their land and into the margins of Argentinian society.

Another target for land-grabbing is Africa. Nine million hectares of land are thought to have been seized there between 2006 and 2009 – about three-quarters of the total amount 'grabbed' worldwide.[9] An international conference on Global Land-Grabbing estimated that more than 80 million hectares (200 million acres) of land were obtained in this way in 2008–09[10] – an area twice the size of California.

The UN predicts that demand for meat will cause the global livestock population to double in size. If those animals are kept in factory farms, even more land will be needed to produce extra feed crops. Some have suggested there isn't the room to keep animals free-range. This is a myth. The entire global population of meat chickens – about 55 billion birds – could be kept free-range on an estate the size of Hawaii.[11]

MONOCULTURE
Soya fields like this one dominate the landscape in large parts of the world.

A worldwide land grab

Over the last 40 years, agricultural land has gained 500 million ha of land from forest and other uses. This is an area **10 x the size of France.**

30% of the world's ice-free surfaces are used to keep or feed livestock.

In the developing world, between 1980 to 2000, more than **100 million ha** of new farmland was created, 80% of it carved out of tropical forests.

The World Bank estimates that up to **240 million ha** of new agricultural land will be needed by 2030 to meet demand.

Poland, Hungary, Romania and Czech Republic:

Western Europeans are buying up land.

Land rush

Food-importing countries are buying up land around the world.

Pakistan:
UAE have bought

324,000 ha

Sudan:
South Koreans have bought

629,000 ha

Tanzania:
Saudi Arabians have bought

500,000 ha

Halving consumption

of meat, dairy and eggs in the EU would result in 23% per capita less use of cropland for food production.

THICKER THAN WATER

An oil-dependent business

Oil and farming are so tightly connected that Albert Bartlett, physics professor at the University of Colorado, described modern agriculture as "using land to convert petroleum into food".[12] Factory farmers are among the oil industry's most important customers.

Whereas traditional farms relied on manual labour, modern agriculture depends heavily on oil and gas-guzzling machinery. It also devours vast quantities of petrochemicals used in fertilisers and pesticides. That's why farmers often take part in fuel protests – oil price rises hit them hard.

One tonne of US maize, a staple feed crop for intensive livestock, takes a barrel of oil to produce.[13] Globally, modern farming methods use an average of two barrels of oil to produce enough fertiliser and pesticide

for one hectare of crops.[14] Yet sources of crude oil and gas are running out, so energy prices are expected to rise.

The global hunger for oil is so great that no potential source seems to be off-limits to energy giants. Places of unspoilt beauty like Northern Alaska, which are already under threat due to climate change, look set to be plundered for the oil that lies there. When animals are farmed on land rather than indoors, they use energy and other resources more efficiently. Organic farming relies much

FINITE RESOURCE
As oil reserves diminish, factory farming will become increasingly expensive.

less on oil. A Cornell University study found that it takes 31 per cent less energy to produce organic maize than conventional crops. Another finding was that if 10 per cent of all US maize were grown organically, it could save the US the equivalent of 4.6 million barrels of oil a year.[15]

The same study also looked at meat production. It found that growing wheat and maize – and probably vegetables, too – is far more energy-efficient than producing meat. Intensive meat production generally uses far more energy than organic meat production. With the exception of organic chicken and eggs, which uses more because the birds live longer and have space to exercise, organic meat production uses up to 38 per cent less energy.[16]

↑ SPACE TO ROAM
If all the meat chickens in the UK were kept free-range, they would only take up the equivalent space of one-third of the Isle of Wight.

INTENSIVE FARMING IS THIRSTY WORK

Fresh water is a precious and limited resource. The UN warns that farming is already by far the dominant cause of water depletion globally.[17] Worldwide, 70 per cent of fresh water is used for agriculture.[18] Around a quarter of this is linked to meat and dairy production.[19]

↑ HIGH SPRINKLERS
Modern sprinkler systems spray fine droplets to mimic natural rainfall.

135

Water footprint

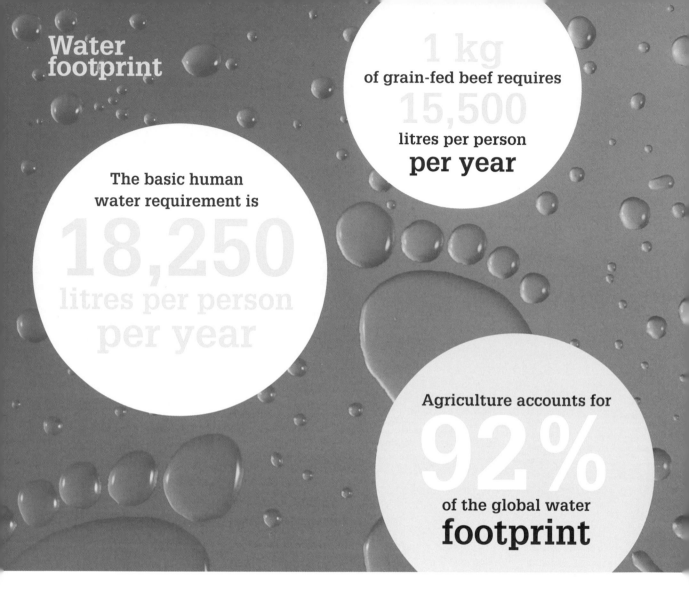

The basic human water requirement is
18,250
litres per person per year

1 kg of grain-fed beef requires **15,500** litres per person **per year**

Agriculture accounts for
92%
of the global water footprint

Rearing animals is a thirsty business. It takes ten times more water to produce meat than vegetables[20], and to get just one kilo of beef, it takes about 90 bathtubs of water.[21] Factory farms are draining lakes and rivers to water crops to feed their animals. Where animals are reared indoors, barns also have to be hosed down and drinking water provided. In contrast, when animals graze, much of the water is rain on grass – a natural process.

Some people say that water can't really be wasted as it's never destroyed. However, it remains a valuable and finite resource. The trouble is, about 97 per cent of it is in the sea, where it's useless unless the salt can be taken out.[22] Water is heavy and expensive to transport, so it's just not possible to move enough of it from places where there's too much to where it's in short supply.

❝ When the well is dry, we know the worth of water. ❞
Benjamin Franklin

Intensive farming at the Al Safi mega-dairy in Saudi Arabia is an example of water wastage. Each cow is said to use 135 litres (30 gallons) of water a day.[23] Here in the Arabian desert, water is sucked up from up to a mile underground is used to drench alfalfa crops for cattle feed. The cows are also kept covered with a fine spray of water all day to keep them cool.

Up to 2 billion people worldwide don't have enough water. Scientists believe that number could reach 7 billion by 2050.[24] At the same time, the UN Food and Agriculture Organisation (FAO) predicts that the livestock population will also double by 2050,[25] most of them in thirsty industrial systems.

If people ate less meat, and more animals were kept in natural grazing systems, it could make a massive difference to water conservation.

Diet	Green, blue and grey water footprint (litres per day)	% reduction as compared with current diet	Blue and grey water footprint (litres per day)	% reduction as compared with current diet
Current diet	4,265		693	
Healthy diet	3,291	23%	557	20%
Combined diet	2,973	30%	512	26%
Vegetarian diet	2,654	38%	467	33%

THE TRUE COST OF FACTORY FARMING

The price of turning grain into meat

People think of factory-farmed meat as cheap, but the true cost of one factory-farmed hamburger has been calculated to be as much as 100 US dollars.[26] If you include energy and water use, the cost of cleaning up pollution and paying for the health problems caused by farms and factories turning grain into meat, as well as the cost of keeping the animal, it all adds up.

Food prices are rocketing worldwide because food production can't keep up with soaring demand, and this is directly linked to factory farming. While extreme weather has also played a part, the colossal demand for cereal crops is adding a massive burden. The pressure of having to supply cereals and soya to feed animals on industrial farms means that prime land can't be used to grow food for people. Industrially reared animals eat up to a third of the world's entire cereal harvest and 90 per cent of its soya.[27] The industry contributes 14.5 per cent of the world's greenhouse gases[28] – creating more greenhouse gas emissions than all our cars, planes and trains put together.

1 ha of cropland produces enough calories to feed

10 people

After animal feed, biofuel and other industrial uses, currently each hectare only feeds

6 people

There is now a dangerous competition between crops for people, crops for industrial farms and crops for cars. If all the human-edible crops used for animal feed were used to feed people directly, we could feed an extra 4 billion people worldwide.[29]

Instead of squandering precious arable crops to feed animals confined indoors, livestock could otherwise be kept on pasturelands or as part of the healthy rotation on a mixed farm. In this way, animals convert something we can't eat – grass – into things that we can:

↑ HARVEST TIME
One third of the world's harvest is destined for livestock. Increasing demand for animal feed puts upward pressure on food prices.

meat, milk and eggs. This is much more efficient than factory farming. Oxfam has warned that increased demand for grain to feed livestock is likely to push food prices "beyond the limits of affordability" for the world's poorest people: "Rises in food prices and pressures on food supplies are likely to be increasingly compounded, perhaps driven, by rising global demand for meat and dairy products."[30]

Most experts agree that the days of food surpluses are over and high prices are here to stay, with potentially catastrophic consequences. Hunger played a key role in the Arab Spring of 2011, with unrest sparked in some places by the high price of bread.[31]

THE TRUE COST

Defenders of factory farming usually claim that it produces affordable meat for the masses. They also say it's wrong that people who can afford higher-welfare meat should question the system. Yet the truth is that factory farming drives up food prices because of the vast amounts of cereals and soya needed to feed the animals. So while intensive farming churns out cheaper, lower-quality meat for people in rich countries, it's at the expense of people living elsewhere.

It's usually the very poorest people in developing countries who pay the real price for bargain meat, often where there's no welfare state to rely on for help. And in the developed world, like Western Europe and the US, how can it be acceptable to expect people on low incomes to have to feed their children on poorer-quality factory farmed food? Yet few governments are willing to tackle the hidden costs of 'cheap' industrial food highlighted by the '100-dollar burger'.

SOY EXPORTS
Soy awaits transport from Brazil to the US, China and Europe. Brazil is the world's largest exporter of soya, much of it grown on cleared forest.

141

GM – FEEDING PEOPLE OR FACTORY FARMS?

The abuse of GM technology

Genetically modified (GM) crops may seem like a quick fix to lots of problems, but they create new ones. The most common versions of GM crops are those that can be doused in weed-killers and are insect-resistant. The pesticides then wipe out almost everything else, even the bugs that help crops.

UK government research has linked GM crops to a dramatic fall in the number of butterflies, bees, weeds and seeds compared with conventional production using ordinary weed-killers and pesticides.[32] A study by the University of California into the ecological impact of GM crops and conventional fertilisers and weed-killers found: "The massive use of transgenic (GM) crops and agrochemical inputs, mainly fertilisers and herbicides … poses grave environmental problems."[33]

Outside the EU, GM crops are commonplace and those most likely to be GM are the ones used in animal feed. The 'big four' GM crops – corn, soya, cotton and rapeseed – are all used for animal feed. GM feed production is highest in the US but the uptake of GM corn and soya is high among developing countries, where factory farming, or growing associated feedstuffs, has taken off.[34]

'Farmers are being sued for having GM crops on their property that they did not buy, do not want, will not use and cannot sell.'
Tom Wiley, Farmer

More and more farmers in developing countries are falling victim to false promises by salesmen from multinational GM companies. Take the tragic case of Shankara Mandaukar, an Indian farmer who struggled for years to make ends meet because of bad weather, weeds and pests. He was easy pickings for a biotech company salesman offering 'magic' seeds with special powers to resist bugs. The price was high but Shankara was desperate and the salesman promised massive yields. After he scraped together the money, the GM seeds failed twice running. Bankrupt and about to lose his farm to loan sharks, he downed a bottle of insecticide and died in agony in the dust outside his home.

DRIVEN TO DESPAIR

Since 1995, more than a quarter of a million Indian farmers have killed themselves. It's believed to be the largest wave of recorded suicides in human history.[35] Few people outside India know about the crisis but the Prince of Wales drew it to public attention in 2008 after he visited the region. He condemned the "truly appalling" death rate and highlighted the role of GM crops in tipping thousands of small farmers over the edge.[36]

Most of the suicides are triggered by money troubles, often involving expensive fertilisers, pesticides, herbicides or GM seeds. Many were cotton farmers, promised bigger profits if they switched to GM seeds. But far from proving pest-proof, a number of these failed as they were wrecked by parasites, leaving farmers with nothing. Saving seeds is fundamental to farming worldwide: seeds are sown, the crops are harvested and the seeds are kept for the next season's planting. But GM multinational Monsanto thinks farmers who save seeds from their GM crops are thieves. The company has even hired private investigators to catch farmers doing it and bring them to court.[37]

To avoid that happening, they have now developed 'terminator seeds' which yield just one crop. But what happens if traditional crops grown in fields bordering these 'terminator' crops become contaminated and are rendered sterile?

GM technology doesn't even seem to reduce food prices. Cereal and soy prices are at a record high, at a time when more GM maize and soy is being grown than ever before.

PADDY FIELD
GM rice with extra vitamin A could save lives in the developing world.

However, there is potential for certain GM developments to have a positive impact on food production. Ingo Potrukus is a German scientist who has spent more than a decade inventing and developing a new type of GM rice to help nourish the world's poorest children.[38]

It's called Golden Rice and has been adapted so that it contains high levels of the nutrient beta-carotene, which the body turns into vitamin A. It looks and tastes the same as ordinary rice, but it has the potential to prevent between one and two million deaths a year in the developing world, and to save as many as 500,000 children from going blind.[39]

VITAMIN DEFICIENCY

Around 124 million people in 118 developing countries suffer from vitamin A deficiency and its potentially deadly side effects. In Southeast Asia alone, more than 90 million children suffer from the condition.[40] By eating just one bowl of Golden Rice a day, there's a decent chance these people could be saved. Yet since 1999, when Potrykus and his collaborator Peter Beyer discovered they could alter rice genes to produce beta-carotene, Golden Rice has been stuck in a laboratory.

It seems odd that Golden Rice has taken so long to reach the table. After all, other GM crops are widespread. In 2008, 13.3 million farmers in 25 countries – 90 per cent of them smallholders in developing countries – were growing GM crops, covering 125 million hectares (300 million acres).[41]

Yet there are genuine concerns that if Golden Rice is approved it could worsen malnutrition problems by discouraging poorer people from taking their own steps to improve their diets by, for example, eating more leafy greens. The WHO argue that there are simpler solutions to vitamin A deficiency. They prefer to promote breastfeeding and supply high-dose vitamin A supplements for deficient children – a solution they say has had "remarkable results".

Despite the delays, Golden Rice now looks set to be approved by governments around the world for production and distribution – and, while they've been waiting, Potrykus and his colleague have developed an even more nutritious version of the rice.

What's particularly attractive about Golden Rice is that it was developed for humanitarian reasons and Potrykus, now in his eighties, remains adamant that there will be no role for profiteering. Throughout the long years in his laboratory and the battle to win approval for Golden Rice, Potrykus was spurred on by the memory of what it's like to go hungry. Growing up in post-war Germany, he and his brother were forced to forage for food. It raises hope that the answers to world hunger can be developed in science labs. Many people believe that if science can solve malnutrition, we have an absolute moral responsibility to let it get to work. Even the Vatican is on-side. The Pope recently issued a statement declaring that governments have a duty to make GM crops more readily available.

CLONING
Pushing animals to extremes

Dolly the sheep was the world's first successfully cloned mammal. Born in 1996 at the Roslin Institute in Scotland, she was the only successful birth out of 277 implanted embryos that led to only 13 pregnancies.

It was Dolly's birth that first inspired the idea of cloning pets. Bankrolled by an eccentric American multi-millionaire, John Sperling,[42] scientists set about trying to clone cats and dogs. Sperling and his partner Joan Hawthorne were fascinated by what the Roslin Institute has achieved and saw an immediate opportunity to use the technique in their own home – to clone their beloved pet dog. Missy, a border collie and husky cross, was getting old, and the couple were prepared to spend whatever it took to immortalise her.

News of the project spread fast and soon they were being contacted by wealthy people from all over the world who wanted to clone their pets. In 2000, they set up a company called Genetic Savings and Clone with a view to starting an international pet-cloning service. Their first focus was on cloning cats, and their funding led to the world's first cloned cat – called Carbon Copy – in 2001. Created and adopted by scientist Duane Kraemer in a Texan laboratory, Carbon Copy hit the headlines a second time in 2011 when she reached her tenth birthday.

Scientists in Korea managed to create the first cloned dog in 2005. By then, Missy had died, but her owners had gene-banked her DNA for future use, so it was flown to Korea and used to create three copies of their late pet.

CLONING LIVESTOCK

In the meantime, Kraemer and his team moved on to other species of clone but their techniques couldn't guarantee the clones would look the same – some had different colouring. Coupled with the vast cost of pet-cloning, this meant the service never really took off. However, cloning scientists found new customers – agricultural animal breeders. They are now developing 'cash cows' that will produce even more meat or milk and, therefore, more money.

Today's scientists can help farmers make copies of their most productive animals – but it comes at a price. The animals they clone are already the supermodels of farming, the result of selective breeding techniques to make them more meaty and milky. But the truth is that cloning copies animals that have already been pushed to their physical limits and are genetically programmed to suffer. That's on top of the serious animal welfare problems linked with the cloning process itself.

The cloning companies make it sound easy. All the customer has to do is send a tiny piece of their chosen animal's flesh

NOT AN EXACT COPY
Environmental development effects mean that the cloned cat Carbon Copy had different fur colour from her genetic mother.

to the company's lab and the scientists do the rest. They put the animal's DNA into eggs whose genetic material has been suctioned out and put the embryos into incubators. A few days later they're transferred to a surrogate mother. Once born, the animals checked over and sent back to the customer.

Sounds simple, but for every cloning success, many more animals are likely to have died.

147

Making Dolly

The cloning process to make Dolly the sheep involved many failed attempts.

Cell from another animal from which the nucleus has been removed

Mammary cell taken from an adult sheep

❶ The cells are fused by an electrical pulse.

Embryo

❷ From 277 cell fusions, 29 early embryos developed.

Sheep

❸ The embryos were implanted into 13 surrogate mothers.

Lamb

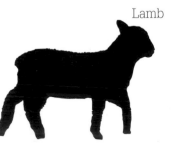

❹ Only one pregnancy went to full term.

Cloning causes a range of horrendous health problems and deformities that are rarely seen otherwise.[43] Despite years of research, the technique hasn't improved much. Surviving pregnancy is the first hurdle, as most die in the womb.[44]

Those that make it to birth are likely to suffer serious health problems including deformed hearts, grossly oversized organs, lowered immune systems and problems with their lungs.

The European Food Safety Authority found that the health and welfare of "a significant proportion" of clones was "adversely affected, often severely and with a fatal outcome".[45] The European Group on Ethics in Science said it saw no "convincing arguments to justify the production of food from clones or their offspring."[46]

PUBLIC RESISTANCE

People don't want animal cloning. Europe-wide polls show nearly two-thirds of people think it's morally wrong. More than half of those surveyed felt cloning for food was unjustifiable. Most said they wouldn't buy meat from clones, and eight out of ten said it should be labelled if it ever became available to buy.[47]

In 2010 there was a public outcry when food from cloned animals slipped unauthorised into the UK food chain. A government inquiry into the incident declared that it wasn't illegal to sell meat or milk from the offspring of cloned animals. The public now has no way of knowing whether the food on their plates has come from cloned animals.

Cloning is just the start. In labs all over the world, scientists are experimenting with animals so that companies can produce more for less. Take meat chickens. In intensive systems, thousands of birds are crammed into sheds. Naturally, they get hot. So scientists have come up with the idea of featherless chickens. Not only do they stay cooler, but they also take up less space, so you can squash more into the same space. Once slaughtered, you don't even have to pluck them. Scientists in China have added human DNA to 300 cows to get them to produce 'human' milk. The idea is to get milk with some of the key properties of human breast milk to help boost an infant's immune system.[48] But can the coming food crisis be fixed in a laboratory or will scientists and big business simply keep pushing already overworked animals to breaking point, risking their welfare and the food they produce?

DOMESTICATED SPECIES
Millennia of selective breeding have already genetically modified domesticated animals.

CUTTING DOWN

Beef production more than **doubled** and chicken meat production went up by a factor of 10 between 1980 and 2002.

At least **75%** of production growth to 2030 is predicted to be in systems that confine animals.

In rich countries, people eat on average

(that's 20% too much).

Meat production in the developing world **tripled** between 1980 and 2002.

Reducing meat, dairy and eggs in the EU by 50% would reduce:

Factor	% reduction
Soybean use as animal feed	76%
Surface and ground water	20%
Cropland use	23%
Nitrogen emissions	40%

Reducing meat intake reduces the risk of type 2 diabetes by 15–42% and cancer by

6–12%

2.1 billion

people worldwide are overweight.

people are malnourished.

Feeding all crops to humans, we could feed an extra

4 billion people

CHAPTER 5
THE SOLUTION

INTRODUCTION

The UN has warned that global food supply needs to grow massively to feed the predicted nine billion people soon expected to live on our planet. Yet today, more than half the food produced worldwide is wasted.

Countries like China and Argentina are now major players in the food industry as they try to meet the soaring demand for meat in the developing world. Their answer is to build it big and churn it out 'cheap'. But those intensive systems cannot feed the world because the grain-feeding of confined animals uses more food than it produces.

There's no doubt that farming must change. To safeguard the future, we have to farm as if tomorrow matters, and that means changing to more sustainable systems. How do we tackle this situation to everyone's benefit, including the animals? To achieve this without factory farming we need a common-sense approach, based on three principles: putting people first, reducing food waste and farming as if tomorrow matters.

Consumer power also has a part to play in changing the future of farming. Buying food from the land, wasting less and cutting down on meat will make a big difference. But what food should we buy to be kinder to animals, to ourselves and to the planet?

HAPPY PIG
Farm animals are kept happy by allowing them to display their natural behaviour. In hot weather, pigs like to wallow in mud to keep cool.

PRINCES, COMMONERS AND SUPERMARKETS: WHERE THE POWER LIES

Forces of change in the food industry

The heir to the British throne, Prince Charles, has long been a champion of sustainable farming. Home Farm, based at the family home of Charles and Camilla, Highgrove House, is a model of sustainability. Completely organic since the mid-1980s, it is home to 180 dairy cows, 150 suckler cows, 130 breeding ewes that produce around 200 lambs a year, and a few rare-breed pigs. It works on a crop-rotation system – a seven-year cycle designed to maximise the richness of the soil.

Home Farm supplies luxury hotels and some products are sold to Duchy Organics, now a partnership with Waitrose supermarket. Although Prince Charles may sell to luxury markets, he doesn't get fancy prices just because of his name. The truth is, Home Farm doesn't always make a profit and some years it struggles to break even. Like most

small farmers, he is having to diversify and is considering making cheese to maximise profit. The revelation that his organic farm struggles despite all the cachet and advantages of his name and status is a sobering reminder of how the odds are stacked against farmers who reject factory farming.

The US and European agricultural systems have been geared towards factory farming since the post-war years. The original motive was to end the years of food shortages and make national food production more self-sufficient, but little thought was given to the long-term consequences. The 1947 Agriculture Act was a defining moment in British farming, kick-starting factory farming in the UK. In the US, the 1993 Farm Bill brought in a package of subsidies for farmers.

Farmers in the UK were encouraged to use the latest chemicals, machines and

VEGETABLE DELICACIES
The author enjoys a vegetarian meal in China.

SUCCESSFUL CAMPAIGN
Campaigners protest against the wasteful slaughter of the Gadhimai festival in Nepal. The practice has now ended.

techniques. Mixed farms were abandoned as farmers began specialising in particular crops and animals. The age-old natural cycle – where crops would be rotated with livestock whose manure replenished tired soil – disappeared and chemical fertilisers were used instead. Farming was now an industry, like producing cars or TV sets, and farmers were told to intensify or get out. Many got on board, some didn't. Just after the Second World War, the UK had around half a million farmers. By the 1980s, numbers had fallen by nearly two-thirds.[1]

Ruth Harrison's 1964 book *Animal Machines* helped raise public awareness of life on factory farms. It described how factory farms revolve entirely round profits and animals are treated like money-making machines.

Meat consumption

On average, over their lifetime, each person in the UK will consume ...

1,400 chickens

24 pigs

19 sheep

Harrison's book was the trigger that would inspire perhaps the greatest campaigning champion for farm animal welfare, Peter Roberts, the founder of Compassion in World Farming (CIWF).

In founding CIWF, Roberts set the wheels in motion for a decades-long struggle against industrial farming. Today CIWF is the world's leading farm animal welfare charity and is an influential force in international discussions on farming. Until 1993, its campaigning was focused mainly on persuading national governments to change legislation in order to improve the lives of farm animals. Since then, the EU has been an important target. This means winning over as many of the 28 member states as possible to influence Europe-wide policies.

Agricultural subsidies are important in the food-production game and work against small producers like Prince Charles. The Common Agricultural Policy (CAP) may sound boring, but it's vital to any debate about factory farming as it underpins the whole system.

The CAP was designed to provide farmers with a fair living and consumers with decent quality food, and also to preserve rural heritage. Yet this complex system has actually encouraged the decline in mixed farming and is partly to blame for animals disappearing from the land. It's now the most expensive and controversial scheme in the EU, costing around £48 billion a year – that's almost half the EU's entire budget. The US has its own version called the Farm Bill. Through this programme, farmers receive billions of dollars of subsidy.[2] The most heavily subsidised US crop is corn (maize), the key feed ingredient of the US 'cheap meat' culture.

27 turkeys

17 ducks

The life's work of **50** laying hens

4 cows

HALF the life's work of a dairy cow

Persuading the EU and other regulators to shift their massive subsidy schemes away from supporting factory farming is a slow process. Although that remains an important focus for campaigners, attention is now turning to businesses, too. Big retailers, supermarkets and fast-food outlets, and the big food-manufacturing companies have a colossal influence on the food system and are capable of making changes more quickly.

CHANGING CHAINS

By working with retailers, it's possible to make a radical difference to the entire food and farming chain, from the way farmers rear their animals to the final product. If retailers decide to make a change, they can do it faster than any government. For example, UK supermarkets Sainsbury's, the Co-op and Marks & Spencer now stock only free-range eggs and the remaining UK retailers have pledged to go cage free on eggs by 2025.

Contrast that with the EU's attempt to ban battery cages. After the EU made the decision in 1999, producers were given a very generous 12 years to switch systems. Yet when the new legislation came into force in 2012, nearly half of the countries were not ready. This meant that tens of millions of hens were still being kept illegally in cages.

When retailers decide what products to sell, consumer views are a powerful lever. Research and polling show that farm animal welfare influences the buying habits of an increasing number of people and, as a result, food companies are more interested than ever in stocking animal-friendly products.

HOW PUTTING PEOPLE

FIRST KEEPS ANIMALS

MIXED FARMING
Cows that graze in fields turn grass into meat, while rotating livestock with crops replenishes the soil.

THE SOLUTIONS

▶ Grazing animals should be kept on pasture as part of the rotation on mixed farms, where crops and livestock move round the farm in harmony. This converts plant life that people can't eat into edible food.

▶ Fish should be fed to people, not livestock. The plundering of our seas through overfishing and feeding them to farm animals must stop.

▶ Strong action in the form of legislation, incentives, research and buying policies is needed from governments, corporations and shoppers alike.

Cereals are a big deal. Worldwide they provide around half the total calories for people, in bread, pasta, tortillas, pies and pizzas. Today, a third of the world's cereal harvest is fed to farm animals.[4] That's enough to feed an extra 3 billion people.[5] Ninety per cent of soy is also used to feed animals, enough to feed a further billion people.[6] It's the same with fish. Up to 30 per cent of the fish landed worldwide are shipped all over the world to feed intensively farmed animals when they could feed people living where the fish are caught.[7]

Industrial farming has thrown farm animals directly into competition with people for food. Cows and sheep turn grass into meat and milk. Chickens and other foraging animals turn foraged food from woodlands and orchards into meat and eggs. Yet we don't let them do it. Instead, they are factory-farmed, taken off the land to be reared indoors on grains or fishmeal grown elsewhere on precious arable land.

Factory farms are food factories in reverse: they waste it, not make it, and squander valuable cropland in the process. For every six kilos of plant protein like cereals fed to livestock, only one kilo of animal protein is given back in the form of food for people.[8]

Cutting the amount of grain fed to farm animals by half would go a long way towards a saner food system. By restoring animals to pasture, it could free up enough grain for more than a billion people. People don't have to choose between eating cereals or meat. Both can be produced far more effectively with the right kind of farming.

❛A third of all the food produced is either binned or left to rot. ❜
United Nations

162

Wasted calories

- 61% of global crop calories are wasted.
- 39% are used for human consumption either directly or as meat or milk.

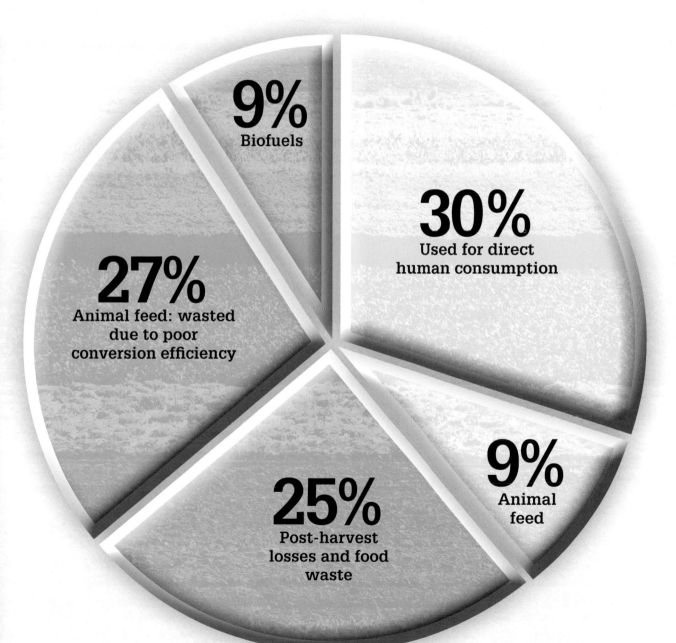

9% Biofuels

30% Used for direct human consumption

27% Animal feed: wasted due to poor conversion efficiency

25% Post-harvest losses and food waste

9% Animal feed

REDUCING FOOD WASTE
Using resources wisely

North America and Europe waste as much as half their food – enough to lift the world's billion hungry people out of starvation up to seven times over.[9] Whether it's from shops or catering companies, or tossed into our own bins at home, food is wasted all along the food supply chain in rich countries.[10] And it's not just fruit and veg: in the UK alone, we waste the equivalent of 50 million chickens every year. The meat waste equivalent of two billion farm animals a year are reared, slaughtered and binned in the EU.

In many countries, food waste ends up in landfill, but space for that is running out. Pure plant-based food waste that's been through strict processing can already be fed to animals, but there's potential to do much more. The Japanese, South Koreans and Taiwanese are all ahead of the game. They've set up food-waste collection and recycling centres to ensure leftovers are properly sterilised and safe for animals reared for meat. If the EU made food recycling easier, the 40 million tonnes of livestock feed imported every year from South America could be slashed and there would be far less pressure on landfill.

Wasted meat

In the UK

1.1–1.5 million pigs | 44–50 million chickens

In UK retail

22,032 tonnes pigmeat | 23,000 pork
= 300,000–417,000 pigs
12240 tonnes beef = 42,000

In UK hospitality

8,769 tonnes pigmeat | 4,329 beef
= 119,00–166,000 pigs | = 15,319
8,769 t chickenmeat = 52,600

Worldwide, the UN says that about a third of food is wasted through being binned or left to rot.[11] It estimates that 28 per cent of the world's agricultural land is used to produce food that's wasted, at a cost of about 750 billion US dollars.[12] An extra billion people could be fed if this were cut by just half.[13]

THE SOLUTIONS

▶ Chickens and other foraging animals should be fed properly treated food waste and allowed to root out food. Kept on mixed farms, they can turn foraged food into eggs and meat.

▶ More should be invested in waste reduction systems. Less food waste should be encouraged at every level – from farmer to corporation to consumer.

GETTING WASTED

Recycling saves energy and resources while creating a cleaner environment. Britain is slowly improving its recycling processes, but we are hugely wasteful.

compared to **11%** in 2001/02

over **40%** of household waste was recycled in England in 2011/12.

of local authority collected waste was sent to landfill in 2011/12 compared to an EU average of 40%.

TOP 5 EUROPEAN RECYCLING COUNTRIES
SWITZERLAND **52%**
AUSTRIA **49.7%**
GERMANY **48%**
NETHERLANDS **46%**
NORWAY **40%**
(with the UK way behind at **17.7%**)

We generate **290m** tonnes of waste every year.

During 2010, regulated waste facilities in England and Wales managed nearly 140m tonnes of waste, processed as follows:

45.9 million tonnes were land filled

41.4 tonnes were moved, before disposal

32.4 tonnes were treated

14.7 tonnes of metal were recycled

5.9 tonnes were incinerated

BIN IT! but always check the recycle label first!

5.3m tonnes of food and drink were sent to landfill in 2013.

1000 years for one plastic bag to fully degrade. Every minute 1m plastic bags are used.

Recycling one tonne of paper uses **70%** less energy than manufacturing from trees and 40% less water.

Hazardous waste in England and Wales has decreased by **30%** since 2004.

2 billion steel cans are recycled in Britain every year. That's enough to circle the Earth **five times**

The UK recycles **17.7%** of its waste.

Waste that cannot be recycled or composted can often be used to generate energy or heat, or to provide residual materials for recycling.

23% of renewable energy comes from waste energy – the equivalent to 726,000 tonnes of oil.

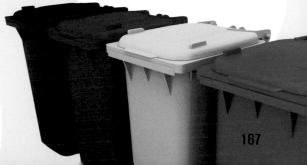

167

FARM AS IF TOMORROW MATTERS
Scaling up best practice

Many farm animals have disappeared from fields and been confined in cramped sheds. This means the 'nutrient cycle' – in which the sun and rain feed the grass, which is fed to animals, whose manure enriches the soil – has been lost. Instead, intensive systems rely heavily on fossil fuel-based fertilisers and chemical pesticides that wreck the soil.

The UN has warned that the productivity of the world's farmland could drop by a quarter this century.[14] Soil erosion already affects almost a third of the world's cropland,[15] and is widespread in the EU.[16]

There is now added pressure from land use for biofuels and the continued growth of industrial livestock. Much greater emphasis is needed on soil-healthy rotational farming with a mix of crops, pastures and farm animals. This relies much less on artificial fertilisers as well as being kinder to animals.

White Oak Pastures in Georgia, USA, provides a glimpse of how future farming could look. Will Harris's family has been raising cattle here since the American Civil War. He farms 1,060 hectares of land and a lot of animals: 1,800 cattle, 50,000 chickens for meat, 1,000 laying hens, 800 sheep and various other species. The farm operates a wholesome type of rotational grazing system. The animals move round the pasture

in succession: big animals followed by smaller animals, followed by poultry. Each of them feeds from the land and returns manure to the land. Working in tune with nature means wildlife thrives too.

It wasn't always this way. After the Second World War, Harris's father started farming industrially. "It was all about pounds of beef produced and nothing about the quality," says Harris. They fed the cattle grain and used hormone implants to make them grow faster. Antibiotics were mixed in feed and pastures doused with chemicals. Harris grew

FREE-RANGE PIGS
Mixed farms rotate animals around pasture to maintain healthy soils.

disenchanted and when he took over the business, he brought it full circle.

It's now a celebrated model of environmental sustainability, animal welfare and good food. His clients include the catering giants Sodexo and the retailers Publix and Whole Foods Market. Could it be done on a big scale? Harris has no doubt. "I know I could scale it up tenfold," he says.

THE SOLUTIONS

▶ **Food should be produced on mixed farms of crops and animals.**
▶ **The vital link between farm animals and the land needs to be restored to improve soil sustainability. Research shows that keeping farm animals outdoors needn't take up huge amounts of space.**[17]

LABELLING MATTERS
Knowing what to buy

The key to consumer choice is knowing what we're buying. But when it comes to shopping, people power is limited by lack of information. Factory farmers don't want shoppers to know the ugly truth of how their meat and eggs are produced. Some people don't want to know, but more and more people do, and that's why clear labelling is so important.

In the EU, eggs now have to be labelled according to how they were produced. Battery eggs, for example, have to be labelled 'eggs from caged hens'. That's not the case for meat and milk. Supermarkets often use misleading labels to make their products seem more animal-friendly than they really are. For example, labels like 'farm fresh', 'country fresh' or 'natural' probably mean it's from a factory farm.

The UK's Red Tractor symbol boasts that it stands for 'choosing high animal welfare standards' but in reality it all too often assures little more than compliance with basic legislation. The Soil Association logo is the gold standard for animal welfare in the UK, though the RSPCA Assured (formerly Freedom Food) scheme also genuinely delivers a higher-welfare choice for people buying meat, milk and eggs.

↓ CHECK THE LABEL
Clear labelling is essential to allow us to know what we are buying.

THE SOLUTIONS

▶ All meat, milk and eggs should be clearly labelled according to how they were produced. This should be mandatory so that everyone can make informed choices.

▶ Labels should be in plain words and there should be guidelines about how they appear so the information can't be buried in tiny print.

FUTURE FOOD
New ways to farm

Science has an important role to play in the menu of the future, and the development of new kinds of foods could help avert the looming food crisis. Some scientists are looking into new types of food that can be produced in ways that are kind to both animals and the planet.

Willem Brandenburg and René Wijffels, at the Dutch University of Wageningen, are exploring the possibility of growing seaweed and algae on a massive scale. Seaweed is easily digested and provides almost as much protein as meat. Dr Brandenburg says he can grow enough seaweed in 360,000 km^2 (140,000 sq miles) to feed the protein requirements of as many as ten billion people. That's an area of sea four times the size of Portugal – which in oceanic terms is tiny. He already has an experimental farm off the Dutch coast of Zeeland and talks enthusiastically about seaweed boosting food supplies without taking over more land. Brandenburg is in no doubt about the scale of the challenge we face. He says: "Over the coming four decades, we'll need to be producing twice as much food with half the inputs; that's why we're looking at plants, plants and plants as the way forward." Wijffels is looking into growing algae as a source of protein and biofuel. He believes it could be far

↑ ALGACULTURE
Algae farms like this one in Indonesia could help boost food production.

more efficient than traditional land-based agriculture and could replace the soya being imported into Europe, largely as animal feed.

Meanwhile, a promising alternative to fish farming is being used in places like Alaska and Japan, and is being explored in other countries. Ocean ranching is where young fish are hatched and reared in captivity before being released into the sea. They then live naturally in the wild before returning to their release point as adults, where they are caught. In 2010, nearly half the commercially caught salmon in Alaska were ocean-ranched[18] and in Japan, at least 90 species have been released either commercially or experimentally.[19] A wide range of other countries, including Scotland, Sweden and Iceland, have actively looked at this form of farming the sea.[20]

Laboratory-produced beef is also being developed. Backed by Microsoft magnate Bill Gates and Google co-founder Sergey Brin,

scientists are working on meat that's made from tiny fleshy strips grown from cow stem cells. In August 2013, the world's most expensive beefburger was cooked and eaten in London in front of the world's press. The experimental burger, made from *in vitro* – laboratory-produced – meat, was the work of Professor Mark Post of Maastricht University and cost about £200,000 to develop.

Gates says it's not about asking everyone to become vegetarian but about creating alternatives and reinventing meat and eggs that are produced in a more sustainable and planet-friendly way. According to New Harvest – an organisation funding research into *in vitro* or 'cultured' meat in the US and Europe – a single cell could produce enough meat to feed the global population for a year. Another benefit is that you can control the amount of fat the new meat contains. It could be made with more healthy fat like omega-3 so it actually helps prevent heart disease.

THE SOLUTION

▶ Governments and businesses should invest more in developing sustainable and welfare-friendly alternatives to meat, milk and eggs.

↓ AQUACULTURE, ZANZIBAR
Harvesting seaweed could help reduce pressure on world fish stocks.

WHAT YOU CAN DO TO HELP
Buying better food

Everyone can play a part in bringing about a better food future. In the developed world, most of us have three great chances a day to help make a kinder, saner food system through the food choices we make. As well as buying the right foods, make sure you don't waste them. This is the simplest way to make a major contribution to a better food system as it helps reduce the amount of land, water and oil used. It also cuts the landfill needed as well as your food bill.

When shopping, buy food from the land, not from factories and try to buy local. Look out for products labelled 'free-range', 'pasture-fed', 'pasture-raised', 'outdoor-reared' or 'organic'. EU laws currently ban the feeding of food waste to farm animals, so organic and free-range chickens and pigs have to be fed on soya and cereals. Until this changes, buying pasture-raised chicken and pork is the most welfare-friendly option.

In the UK, look out for organic and RSPCA Assured labels and in the Netherlands buy from the Beter Leven (Better Life) scheme.

In the US, the Animal Welfare Assured (AWA) and Global Animal Partnership (GAP) labels are best. In Australia a good option is anything with the 'RSPCA approved farming' logo. Another small change you can make is to eat less meat. Cutting down – going meat-free on Mondays, for example – is a simple step toward avoiding Farmageddon.

WHAT TO BUY

LAMB

Sheep usually spend more time outdoors than most other farm animals so products from lamb or sheep are a good choice.

BEEF

Choose pasture-reared, pasture-fed, grass-finished or organic beef as these allow animals to express their instinctive behaviours and feed in a more natural way.

DAIRY

Choose organic, or milk, cheese and yoghurts produced under one of the dedicated animal welfare schemes like RSPCA Assured. Whilst dairy cows in Britain graze in fields in the summertime, cows in the US and other animals are kept indoors, so to avoid this choose AWA standards or buy organic.

EGGS

Buy free-range or organic eggs as these hens will have been given access to the outdoors. In the EU, eggs must now be labelled according to how they were produced. Elsewhere, if the label doesn't say that the hens were given outdoor access, that means they're probably from battery-caged hens. Other foods such as mayonnaise, cakes and pasta contain eggs, so unless the label says they're 'cage-free' or free-range, they're likely to be from caged hens.

RSPCA ASSURED

‘ Our vision is for all farm animals to have a good life. ’

RSPCA

CHICKEN MEAT

Free-range and organic chicken is not only kinder, it contains up to 50 per cent less fat than its factory-farmed equivalent. Most supermarket chickens are kept indoors their whole lives whereas free-range, pasture-raised or organic chickens will have had access to the outdoors. In the US, look for Animal Welfare Approved (AWA), Global Animal Partnership (GAP), or Certified Humane. Avoid labels like 'farm fresh' or 'corn-fed'.

TURKEY MEAT

Choose free-range and organic and, in the UK, look out for the RSPCA Assured label. Slow-growing turkey breeds like Norfolk Black, Black Wing Bronze and Cambridge Bronze are also a more compassionate choice.

↑ MEAT MATTERS
Buying high-welfare, high-quality meat while eating meat less often is better for the planet and better for your health.

FARMED FISH

This is tricky as many wild fish species are threatened. Avoid carnivorous species of farmed fish, such as salmon, trout, cod and halibut, as these are likely raised on feed made from wild fish. If sustainably caught, wild salmon and trout is a better option. With farmed salmon from Scotland, guaranteeing any is seal-friendly, where seals are not shot as part of predator control, is difficult. However, if you do buy farmed fish, look out for those produced under an organic scheme.

PORK, BACON AND SAUSAGES

In the EU, look for free-range or organic, where the pigs are born and reared in systems with outdoor space. 'Outdoor bred' means the pigs are born outside but then reared indoors. 'Outdoor reared' is a better choice, meaning the pigs have spent much of their lives outdoors. In the US, avoid pork or bacon from systems that confine pregnant sows – buy from AWA, Certified Humane, the GAP 5-step Animal Welfare Rating Program, Organic and American Humane Certified.

Avoiding Farmageddon is easy. Simple steps like eating what we buy and eating less and better meat can make a difference. When we choose alternatives to industrial factory farming – like free-range, pasture-raised or organic – supermarkets and policymakers take note. Things begin to change – from Farmageddon to a better future for people, animals and the planet.

COMPASSION IN WORLD FARMING
Campaigning to end animal cruelty

Compassion in World Farming (CIWF) is the world's leading farm animal welfare charity. Founded 50 years ago, it now spearheads a global movement of people worried about how factory farming mistreats animals, wastes precious resources and fails to meet the needs of the public.

CIWF was launched in 1967 by Hampshire dairy farmer Peter Roberts and his wife Anna, who didn't like the way post-war farming was changing. They were concerned about the rise of factory farming and how it was growing apart from the wellbeing of animals and the environment.

Peter and Anna believed the world should be fed humanely and sustainably and in a more natural way. Horrified by the cage and crate systems being used, Peter took his concerns to the big animal charities of the day. When they did little to help, he called a meeting with friends around his kitchen table and took the bold step of founding CIWF. He was a man of intense integrity and humility and used his considerable skills as a speaker and campaigner to influence change.

From its humble beginnings, CIWF has become an influential force in international discussions wherever farm animal welfare is

TIRELESS CAMPAIGNER
CIWF founder Peter Roberts during an early campaign against factory farming.

at stake. Our views are sought by policy-makers and governments around the globe – whether it's in Europe, China, South Africa or the United States. CIWF's award-winning undercover investigations have exposed the reality of modern factory farms and brought the plight of farm animals to the attention of the world's media. Through our hard-hitting campaigns and political lobbying, we have secured many landmark changes that have improved the lives of billions of farm animals around the world.

Our campaigns have already seen the beginning of the end for many of the worst aspects of factory farming. The EU has listened to our voice and enacted legislation on a number of issues. In 1987 the UK government voted to phase out cruel veal crates following a court case and campaign from CIWF. The ban came into force in the UK in 1990. Six years later, continued pressure from us also resulted in legislation to ban veal crates across Europe from 2007.

Because of this campaign, which involved many activities with our thousands of supporters, veal crates are now outlawed in all countries in the EU.

Following intense campaigning by Compassion, keeping pregnant sows in narrow stalls has been banned in the UK since 1999. Continued pressure has meant that, since January 2013, these stalls are now prohibited across the EU after the first four weeks of pregnancy.

In 1999, against all the odds, the EU agreed to ban barren battery cages for laying hens from 2012. It was hailed by many as the single biggest victory for animal welfare in recent history, and we are now working tirelessly to ensure the ban is enforced across the whole EU.

In 2005, the export subsidies paid to farmers transporting live cattle to countries outside the EU were scrapped. We will continue to fight for an end to all live exports from the EU.

In 1997, animals were legally recognised by the EU as sentient beings capable of feeling pain and discomfort, following a ten-year-long campaign. This fundamental agreement now underpins and paves the way for all future improvements to farm animal welfare in Europe.

Since 2007, CIWF's Food Business team has been working with some of the world's biggest food companies – retailers, producers, manufacturers and food service companies – to place farm animal welfare at the forefront of their corporate social responsibility agendas. CIWF works with companies like McDonald's and Unilever – key to our drive towards a more ethical and sustainable food supply. More than 750 million animals are already set to benefit every year thanks to our work with more than 700 companies.

None of these achievements would have been possible without the unstinting support and actions of our thousands of dedicated supporters – individuals all over the world who feel passionately about animal welfare and want to make a real difference. Working together, we can make genuine and lasting changes that positively impact animals. Our campaigns only work because governments, retailers, farmers and businesses know that we have the backing of so many ordinary people who are willing to stand up and be counted to help animals.

Over the next five years, we're launching a 21st-century agricultural revolution to end all forms of cruelty associated with 'modern' intensive farming. We want a kinder, safer, fairer model of farming that works for animals, people and the planet.

Your support will make all the difference.

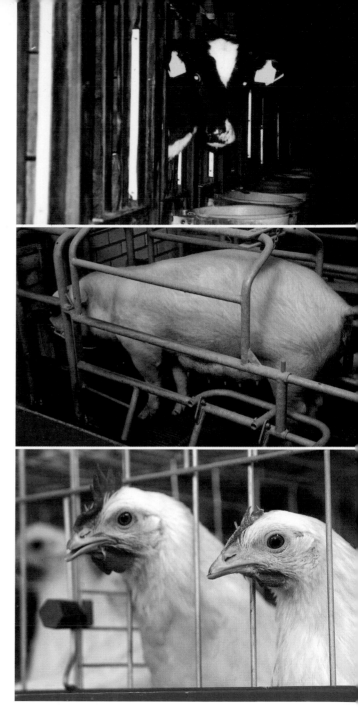

UNNECESSARY CRUELTY
CIWF campaigns to end cruel intensive farming practices such as veal crates (top), sow stalls (centre) and caged hens (bottom).

REFERENCES

Introduction

Defra, *Agriculture in the UK 2010*.

Government Office for Science, *Foresight Project on Global Food and Farming Futures Synthesis Report C1: Trends in food demand and production*, 2011; S. Msangi and M. Rosegrant, *World agriculture in a dynamically-changing environment: IFPRI'S long term outlook for food and agriculture under additional demand and constraints*, paper written in support of Expert Meeting on 'How to Feed the World in 2050', Rome, FAO, 2009; H. Steinfeld et al., *Livestock's Long Shadow, environmental issues and options*, FAO, Rome, 2006, Introduction, p. 12.

FAO, *State of the World Fisheries and Aquaculture 2010*, UN Food and Agriculture Organization, Rome.

Calculated from FAOSTAT online figures for global grain harvest (2009) and food value of cereals. Based on a calorific intake of 2,500 kcalories per person per day.

WHO press release, 'World Health Day 2011, Urgent action necessary to safeguard drug treatments', 6 April 2011, http://www.who.int/media-centre/news/releases/2011/whd_20110406/en/index.html.

Chapter 1 Nature

Rachel Carson, *Silent Spring*, Penguin, London, 1962 (2000 reprint).

http://www.esri.com/mapmuseum/mapbook_gallery/volume23/agricultureI.html (accessed 13 July 2012).

http://nass.usda.gov/Statistics_by_State/California/Publications/California_Ag_Statistics/2010cas-ovw.pdf (accessed 13 July 2012).

Calculated from formula: 200 dairy cows produce as much manure as a town of 10,000 people: *Animal waste pollution in America: an emerging national problem, 1997. Environmental risks of livestock and poultry production*. A report by the Minority Staff of the US Senate Committee on Agriculture, Nutrition and Forestry for Senator Tom Harkin.

http://www.farmland.org/programs/states/futureisnow/default.asp (accessed 13 July 2012).

http://www.sraproject.org/wp-content/uploads/2007/12/dairytalkingpoints.pdf

British Trust for Ornithology (BTO), 'Breeding birds in the wider countryside 2010. Trends in numbers and breeding performance for UK birds' (accessed July 2011) http://www.bto.org/about-birds/bird-trends; British Trust for Ornithology (BTO), 'Breeding birds in the wider countryside, Trends in numbers and breeding performance of UK birds. Section 4.2, Latest long term alerts', http://www.bto.org/birdtrends2010/discussion42.shtml

Defra, 'Wild bird populations: farmland birds in England 2009', news release 29 July 2010.

Defra statistical release, 'Wild bird populations in the UK [to 2009]', 20 January 2011, http://archive.defra.gov.uk/evidence/statistics/environment/wildlife/download/pdf/110120-stats-wild-bird-populations-uk.pdf.

http://www.chesapeakebay.net/issues/issue/agriculture#inline (accessed 7 August 2012).

US EPA, Chesapeake Bay Program, 'Health of Freshwater Streams in the Chesapeake Bay Watershed', www.chesapeakebay.net/status_stream-health.aspx?menuitem=50423.

Bumblebee Conservation Trust, www.bumblebeeconservation.org.uk (accessed July 2011).

C. Carvell et al., 'Comparing the efficacy of agri-environment schemes to enhance bumble bee abundance and diversity on arable field margins', *Journal of Applied Ecology* (2007), 44, pp. 29–40.

'Colony Collapse Disorder and the Human Bee', 12 August 2008, http://www.articlesbase.com/environment-articles/colony-collapse-disorder-and-the-human-bee-519377.html (accessed 21 May 2013).

Los Angeles Times, 'Pesticides suspected in mass die-off of bees', 29 March 2012, http://articles.latimes.com/2012/mar/29/science/la-sci-bees-pesticides-20120330 (accessed 21 May 2013); *Natural News*, 'Confirmed: Common pesticide crashing honeybee populations around the world', 10 April 2012, http://www.naturalnews.com/035518_honey_bees_pesticides_science.html (accessed 21 May 2013).

D. Goulson, University of California, 'David Goulson: Ecology and Conservation of Bumble Bees', 17 April 2013, http://entomology.ucdavis.edu/News/David_Goulson___Ecology_and_Conservation_of_Bumble_Bees/ (accessed 21 May 2013).

'Colony Collapse Disorder and the Human Bee', 12 August 2008, http://www.articlesbase.com/environment-articles/colony-collapse-disorder-and-the-human-bee-519377.html (accessed 21 May 2013).

A. Benjamin and B. McCullum, *A World Without Bees*, Guardian Books, London, 2008.

Los Angeles Times, 'Hives for hire', 3 March 2012.

A. Mood, *Worse things happen at sea: the welfare of wild-caught fish*, 2010, www.fishcount.org.uk.

A. G. J. Tacon and M. Metian, 'Global overview on the use of fish meal and fish oil in industrially compounded aquafeeds: Trends and future prospects', *Aquaculture*, 285 (2008), pp. 146–58.

R. L. Naylor et al., 'Feeding aquaculture in an era of finite resources', *PNAS*, 106(36) (2009), pp.15103–10; Tacon and Metian, 'Global overview'.

FAO, *State of the World Fisheries and Aquaculture*, 2010.

M. J. Costello, 'How sea lice from salmon farms may cause wild salmonid declines in Europe and North America and be a threat to fishes elsewhere', *Proc. R. Soc. B*, 276 (2009), pp. 3385–94; M. Krkošek et al., 'Epizootics of wild fish induced by farm fish', PNAS, 103(42) (2008), pp. 15506–10; M. Krkošek et al., 'Sea lice and salmon population dynamics; effects of exposure time for migratory fish', *Proc. R. Soc. B*, 276 (2009), pp. 2819–28.

SEPA, *Regulation and monitoring of marine caged fish farming in Scotland*, Annex H, Scottish Environment Protection Agency, May 2005.

M. Krkošek et al., 'Declining Wild Salmon Populations in Relation to Parasites from Farm Salmon', *Science*, vol. 318 no. 5857 (2007), pp. 1772–5.

Orr, D. 'The Way to Save the Salmon', *Independent*, 30 July 1999.

USDA Nutrient Database, 2011, USDA Agricultural Research Service, National Agricultural Library, Nutrient Database for Standard Reference, Release 24, Nutrient Data Laboratory Home Page: http://ndb.nal.usda.gov/ndb/foods/list (accessed 2 October 2012).

R. A. Hites et al., 'Global Assessment of Organic Contaminants in Farmed Salmon', *Science*, vol. 303 no. 5655 (2004), pp. 226–9, http://www.sciencemag.org/content/303/5655/226.short.

Animal Aid, 'The "humane" slaughter myth', http://www.animalaid.org.uk/h/n/CAMPAIGNS/slaughter/ (accessed 13 August 2013).

Compassion in World Farming, 'Suffering at slaughter exposed by new film', http://ciwf.org.uk/news/factory_farming/suffering_exposed_by_film.aspx (accessed 13 August 2013).

http://thinkexist.com/quotation/i-hope-to-make-people-realise-how-totally/380118.html (accessed 13 August 2012).

Chapter 2 Health

Hansard, 13 May 1953, 1327–43.

Chief Medical Officer's Annual Report, 2008, chapter: 'Antimicrobial resistance: up against the ropes'.

Dr Margaret Chan, Director-General World Health Organization (WHO), speaking on World Health Day, 7 April 2011.

Committee for Medicinal Products for Veterinary Use, *Reflection Paper on the Use of Fluoroquinolones in Food-producing Animals in the European Union: Development of Resistance and Impact on Human and Animal Health*, 2006, www.emea.europa.eu/pdfs/vet/srwp/18465105en.pdf.

C. Nunan and R. Young, *MRSA in farm animals and meat: a new threat to human health*, Soil Association, 2007.

Ibid

L. Garcia-Alvarez et al., 'Meticillin-resistant *Staphylococcus aureus* with a novel *mecA* homologue emerging in human and bovine populations in the UK and Denmark: a descriptive study', *Lancet Infectious Diseases*, 2011.

EFSA-ECDC, 'The European Union Summary Report on Trends and Sources of Zoonoses, Zoonotic Agents and Food-borne Outbreaks in 2009', *EFSA Journal*, 2011, 9(3), 2090.

WHO (2010) Cumulative Number of Confirmed Human Cases of Avian Influenza A/(H5N1), reported to WHO, 9 August 2011, www. who.int/csr/disease/avian_influenza/country/cases_table_2011_08_09/en/index.html.

M. Du Ry van Beest Holle, 'Human-to-human transmission of avian influenza A/H7N7, The Netherlands, 2003', *Eurosurveillance* 10(12), 1 December 2005, pp. 264–8; http://www.eurosurveillance.org/em/v10n12/1012-222.asp.

D. MacKenzie, 'Five easy mutations to make bird flu a lethal pandemic', *New Scientist*, 24 September 2011, p.14 (online article 21 September).

C. J. L. Murray et al., 'Estimation of potential global pandemic influenza mortality on the basis of vital registry data from the 1918–20 pandemic: a quantitative analysis', *Lancet*, 368 (2006), pp. 2211–18.

World Health Organization, 2010, http://who.int/csr/don/2010_05_14/en/index.html (accessed 30 May 2012).

GCM website, http://www.granjascarroll.com/ing/ing_historia.php (accessed December 2011).

S. M. Burns, 'H1N1 Influenza is here', *Journal of Hospital Infection*, 17 July 2009, http//download.thelancet.com/flatcontentassets/H1N1-flu/epidemiology/epidemiology-76.pdf (accessed 27 July 2012).

Oxford Centre for Animal Ethics, news release, 'Bird Flu Will Remain a Threat as Long as Factory Farms Exist', 17 February 2012, http://www.oxfordanimalethics.com/2012/02/news-release-bird-flu-will-remain-a-threat-as-long-as-factory-farms-exist/.

BBC News, 'Chinese baby milk scare "severe"', 13 September 2008, http://news.bbc.co.uk/1/hi/world/asia-pacific/7614083.stm (accessed 18 July 2012).

Clenbuterol side effects website, http://www.clenbuterolsideeffects.org/ (accessed 4 October 2012); Clenbuterol website, http://www.lenbuterol.tv/clenbuterol-side-effects/ (accessed 4 October 2012); *Independent*, 'Clenbuterol: The new weight-loss wonder drug gripping planet zero', 20 March 2007, http://www.independent.co.uk/life-style/health-and-families/health-news/clenbuterol-the-new-weightloss-wonder-drug-gripping-planet-zero-441059.html (accessed 4 October 2012).

People's Daily, 'Three arrested in pig meat food poisoning of 300 people in Shanghai', 4 November 2006, http://english.peopledaily.com.cn/200611/04/eng20061104_318172.html (accessed 18 July 2013).

China Daily, 'Wen urges cleanup of algae-stenched lakes', 1 July 2007, http://www.chinadaily.com/cn/china/2007-07/01/content_907145.htm (accessed 4 October 2012).

'Development of organic agriculture in Taihu Lake region governance of agricultural nonpoint source pollution', http://eng.hi138.com/?i274195_Development_of_organic_agriculture_in_Taihu_Lake_region_governance_of_agricultural_nonpoint_source_pollution (accessed 23 July 2012). An Olympic-sized swimming pool holds 2,500 cubic metres of water.

iWatch News, '"Free for all" decimates fish stocks in the South Pacific', http://www.iwatchnewsorg./2012/01/25/7900/free-all-decimates-fish-stocks-southern-pacific (accessed 3 August 2012).

Institut de recherche pour le développement, 'Scientists working for responsible fishing in Peru', scientific bulletin no. 349 – May 2010, http://www.en.ird.fr/content/download/17178/146692/.../4/.../FAS349a-web.pdf; C. E. Paredes, 'Reforming the Peruvian anchoveta sector', Instituto del Peru, July 2010, http://www.ebcd.org/pdf/presentation/164-Paredes.pdf; Y. Evans and S. Tveteras, *Status of Fisheries and Aquaculture Development in Peru: Case Studies of Peruvian Anchovy Fishery, Shrimp Aquaculture, Trout and Scallop Aquaculture.* FAO, Rome, 2011, available from www.fao.org/.

Seafish, 'Seafish publishes comprehensive review of feed fish stocks used to produce fishmeal and fish oil for the UK market', 16 April 2012, http://www.seafish.org/about-seafish/news/seafish-publishes-comprehensive-review-of-feed-fish-stocks-used-to-produce-fishmeal-and-fish-oil-for-the-uk-market (accessed 3 August 2012).

Seafish, *Fishmeal and Fish Oil Facts and Figures*, 2011, www.seafish.org.

Evans and Tveteras, *Status of Fisheries and Aquaculture Development in Peru.*

W. H. Dietz, 'Reversing the tide of obesity', *Lancet*, 378 (2011), pp. 744–6.

Jon Ungoed-Thomas, '"Healthy" chicken piles on the fat', *Times*, 3 April 2005, http://www.timesonline.co.uk/.

A. P. Simopoulos, 'The importance of the omega-6/omega-3 fatty acid ratio in cardiovascular disease and other chronic diseases', *Experimental Biology and Medicine*, 233 (2008), pp. 674–88.

C. A. Daley et al., 'A Literature Review of the Value-Added Nutrients found in Grass-fed Beef Products', June 2005, draft manuscript available at All Things Grass Fed: A cooperative project between California State University, Chico College of Agriculture and University of California Cooperative Extension, http://ww.csuchico.edu/grassfedbeef/; A. P. Simopoulos (2000), 'Human requirement for N-3 polyunsaturated fatty acids, Poultry Science, 79(7) (2000), pp. 961–70; A.P. Simopoulos, 'The Importance of the Omega-6/Omega-3 Fatty Acid Ratio in Cardiovascular Disease and Other Chronic Diseases'.

H. Pickett, 'Nutritional benefits of higher welfare animal products', 2012, http://www.ciwf.org.uk/includes/documents/cm_docs/2012/n/nutritional_benefits_of_higher_welfare_animal_products_report_june2012.pdf .

J, D, Wood et al., 'Fat deposition, fatty acid composition and meat quality: A review', *Meat Science*, 78 (2008), pp. 343 –58.

H. Pickett, 'Nutritional benefits of higher welfare animal products'.

http://meatonomics.com/2013/08/15/each-time-mcdonalds-sells-a-big-mac-were-out-7/

Anthony J. McMichael et al., 'Food, livestock production, energy, climate change, and health', *Lancet*, vol. 370, issue 9594 (2007), pp. 1253–63.

Earth Policy Institute, 'Peak Meat: US Meat Consumption Falling', Source: EPI from USDA, US Census, http://www.earth-policy.org/data_highlights/2012/highlights25, (accessed 21 October 2015).

Meat &Poultry staff, 'Flexitarians will increase in 2012: study', M&P news online, 28 December 2011, http://www.meatpoultry.com/News.

S. Heliez et al., 'Risk factors of new Aujeszky's disease virus infection in swine herds in Brittany (France)', *Veterinary Research*, 31 (2000), pp. 146–7, http://www.vetres.org/ (accessed 6 August 2012).

BPEX Weekly, 'French pig producers will go bust', 3 December 2010, http://www.bpex.org/bpexWeekly/BW031210.aspx (accessed 12 August 2012).

Independent, 'Farmers and greens fight the war of the killer seaweed', 15 August 2011, http://www.independent.co.uk/environment/nature/farmers-and-greens-fight-the-war-of-the-killer-seaweed-2337803.html (accessed 6 August 2012); *The Connexion*, 'Brittany beaches after toxic fumes', 1 September 2001, http://www.connexionfrance.com/50-brittany-beaches-closed-after-toxic-fumes-kill-boar-13715-view-article.html (accessed 6 August 2012); *Daily Mail*, 'Holidaymakers warned of deadly seaweed on Brittany's popular beaches', 28 July 2011, http://www.dailymail.co.uk/travel/article-2019700/Brittany-seaweed-warning-Holidaymakers-told-beware-toxic-fumes-rotting-seaweed.html (accessed 6 August 2012); *Daily Telegraph*, 'Toxic seaweed on French coast sparks health fears', 22 July 2011, http://www.telegraph.co.uk/news/worldnews/europe/france/8655329/Toxic-seaweed-on-French-coast-sparks-health-fears.html (accessed 6 August 2012); *Guardian*, 'Brittany beaches hit by toxic algae', 27 July 2011, http://www.guardian.co.uk/environment/2011/jul/27/brittany-beaches-toxic-algae-boars (accessed 6 August 2012); *The Horse*, 'Horse dies in Decomposing Seaweed; Toxic Gas Blamed', 6 August 2009, http://www.thehorse.com/ViewArticle.aspx?ID=14674 (accessed 6 August 2012); *Guardian*, http://www.guardian.co.uk/world/2009/aug/10/france-brittany-coast-seaweed-algae (accessed 6 August 2012); *Science Ray*, 'Green algae is fatal to men', 11 September 2011, http://scienceray.com/technology/green-algae-is-fatal-to-men/ (accessed 6 August 2012).

Food and Water Watch website, http://www.factoryfarmingmap.org/facts/ (accessed 21 May 2013)

J. Trotter, 'Hogwashed', *Waterkeeper* magazine, Summer 2004, http://www.waterkeeper.org/ht/a/GetDocumentAction/i/9899 (accessed 6 August 2012).

J. Tietz, 'Boss Hog', *Rolling Stone*, 14 December 2006, http://regional-workbench.org/USP2/pdf_files/pigs.pdf (accessed 6 August 2012).

Rodale Institute, extracts from Senate Testimony by Rick Dove, 2002, http://newfarm.rodaleinstitute.org/depts/pig_page/rick_dove/index.shtml (accessed 6 August 2012); *Waterkeeper* magazine, Summer 2004, http://www.waterkeeper.org/ht/a/GetDocumentAction/i/9899 (accessed 6 August, 2012); *Waterkeeper* Magazine, Fall 2005, http://www.waterkeeper.org/ht/a/GetDocumentAction/i/9903 (accessed 6 August 2012).

Bracke, M. B. M., Spruijt, B. M. Metz, J. H. M. 1999. Overall animal welfare assessment reviewed. Part 1: Is it possible? Neth. J. Agri. Sci. 47, 279–291.

Fraser, A. F. Broom, D. M. 1990. Farm Animal Behaviour and Welfare, 3rd edition. CAB International, Wallingford, Oxon. Duncan, I. J. H. 1996. Animal welfare defined in terms of feelings. Acta Agricultureae Scandinavica, Sect. A., Anim. Sci. Suppl. 27, 29–35. Lawrence, A. B. 2007. What is Animal Welfare? In: Branson, E. Fish Welfare. Blackwell Publishing, Oxford.

Farm Animal Welfare Council, http://webarchive. nationalarchives.gov.uk/20121007104210/http:/www.fawc. org.uk/freedoms.htm

European Union, 2015. Eurobarometer Special Barometer 442 Attitudes of Europeans Towards Animal Welfare.

Koba, Y., H. Tanida. 2001. App. Anim. Behav. Sci. 73, 45–58.

Munksgaard, L., Jensen, M. B., Pedersen, L. J., Hansen, S.W., Matthews, L. 2005. App. Anim. Behav. Sci. 92, 3–14. Fraser, A. F., Broom, D. M. 1997. Farm Animal Behaviour and Welfare. CABI Publishing. Chapter 8.

Petherick, J. C., Waddington, D., Duncan, I. J. H. 1990. Behav. Proc. 22, 213–26. Weeks, C. A., Nicol, C. J. 2006. World Poultry Sci. J. 62, 296–307.

2010 FAOSTAT:http://faostat.fao.org; FAO, Livestock's Long Shadow: Environmental Issues and Options, Rome, 2006 – proportion industrially reared.

Numbers indicate broiler chickens sold and farms with broiler chicken sales, taken from the USDA Census of Agriculture through 2007, Census information, www. agcensus.usda.goc/Publications/2007/Full_Report/ Volume_1,_Chapter_1_US/st99_I_001_001.pdf.

USDA, National Agriculture Statistics Service, 2010, 'Broilers: Inventory by State, US' http://www.nass.usda.gov/ Charts_and_Maps/Poultry/brlmap.asp (accessed 1 December 2011); United States Department of Agriculture, 2007 Census of Agriculture – Georgia, http://www.agcensus. usda.gov/Publications/2007/Full _Report/Volume_1,_ Chapter_1_State_Level/Georgia/gav1.pdf (accessed 16b November 2011).

FAOSTAT data, 2010, http://faostat.ffao.org

The New Georgia Encyclopedia, http://www. georgiaencyclopedia.org/nge/Article.jsp?id=h-1811 (accessed 1 December 2011).

Knowles T. G., Kestin S. C., Haslam S. M., Brown S. N., Green L. E., et al. (2008) 'Leg Disorders in Broiler Chickens: Prevalence, Risk Factors and Prevention' PLoS ONE 3(2):e1545, 10.1371/journal. Pone. 0001545. http://www. plosone.org/article/info

Pilgrim's Pride, http://turnaround.org/cmaextras/ PilgrimsPride.pdf (accessed 16 December 2011).

D. L. Cunningham, 'Cash Flow Estimates for Contract Broiler Production in Georgia: A 30-Year Analysis', The University of Georgia College of Agricultural and Environmental Sciences, 31 January 2011, http://www.caes. uga.edu/Publications/pubDetail.cfm?pk_id=7052 (accessed 5 December 2011).

Interview conducted by Compassion in World Farming of Southern Poverty Law Center members who have worked with catchers, 2 November 2011.

Bureau of Labor Statistics, Table SNR12, 'Highest incidence rates of total nonfatal occupational illness cases 2010', Bureau of Labor Statistics US Deparment of Labor, October 2011.

Wage and Hour Division, US Department of Labor, Poultry Processing Compliance Survey Fact Sheet, US Department of Labor, 2001.

C. Vicente, GRAIN, Buenos Aires, personal communication, 2012.

Soybean and Corn Advisor website, http://www. soybeansandcorn.com/Argentina-Crop-Acreage (accessed 21, May 2013).

Chicago Tribune, 'ANALYSIS – Argentina's soy addiction comes back to bite farmers', 22 April 2013, http://articles. chicagotribune.com/2013-04-22/news/sns-rt-argentina-soy-analysisl2nod6135-20130422_1_soy-yields-corn-yields-pampas-farm-belt (accessed 21 May 2013).

Defra press release, 'Vince Cable signs multi-million-pound export deal to China', 8 November 2010, http://www. defra.gov.uk/news/2010/11/08/export-pig-china/ (accessed 18 July 2012).

International Finance Corporation, 'Muyuan Pig. Summary of proposed investment', 2010, http://www.ifc. org/ifcext/spiwebsite1.nsf/Project-Display/SPI_DP29089 (accessed 18 July 2012).

Thomas K. Rudel et al., 'Agricultural intensification and changes in cultivated areas, 1970–2005', PNAS, 106 (49) (2009), pp. 20675–80.

Calculation by Compassion in World Farming, 2011.

J. Lundqvist, C. de Fraiture and D. Molden, Saving Water: From Field to Fork – Curbing Losses and Wastage in the Food Chain, SIWI Policy Brief, 2008.

R. K. Pachauri, 'Global warning! The impact of meat production and consumption on climate change', CIWF Peter Roberts Memorial Lecture, London, 8 September 2008, http://www.ciwf.org.uk/includes/documents/cm_ docs/2008/l/I_London_08septo8.pps.

C. Vincente, GRAIN, Buenos Aires, personal communication, 2012.

Soybean and Corn Advisor website, http://www. soybeansandcorn.com/Argentina-Crop-Acreage (accessed 21 May 2013).

Chicago Tribune, 'ANALYSIS – Argentina's soy addiction comes back to bite farmers', 22 April 2013, http://articles. chicagotribune.com/2013-04-22/news/sns-rt-argentina-soy-analysisl2nod6135-20130422_1_soy-yields-corn-yields-pampas-farm-belt (accessed 21 May 2013).

BBC, 'Argentina's forest people suffer neglect', http://news. bbc.co.uk/1/hi/programmes/from_our_own_ correspondent/7014197.stm (accessed 21 October 2015).

9 D. Headley, S. Malaiyandi and F. Shenggen, *Navigating the Perfect Storm: Reflections on the Food, Energy and Financial Crises*, August 2009, IFPRI (Online Resource) available at http://www.ifpri.org/sites/default/files/publications/ifpridpoo889.pdf.

10 'Global Land Grabbing: Update from the International Conference on Global Land Grabbing', ISS, 2011, http://www.iss.nl/fileadmin/ASSETS/iss/Documents/Conference_programmes/LDPI_conference_summary_May_2011.pdf (accessed 20 August 2012).

11 Calculation by Compassion in World Farming, 2012.

12 A. A. Bartlett, 'Forgotten Fundamentals of the Energy Crisis', 1978, http://www.npg.org/specialreports/bartlett_section3.htm (accessed 7 September 2012).

13 World *Bank, World Development Report 2008. Agriculture for Development, 2008*, chapter 2, 'Agriculture's performance, diversity and uncertainties', http://siteresources.worldbank.org.

14 FAO, *The energy and agriculture nexus*, Environment and natural resources working paper 4. Rome, 2000, chapter 2, 'Energy for agriculture', http://www.fao.org/.

15 Pimentel, *Impacts of Organic Farming*.

16 Soil Association, *Energy efficiency of organic farming: analysis of data from existing Defra studies*, published 31 January 2007.

17 H. Steinfeld et al., *Livestock's Long Shadow: environmental issues and options*, chapter 4, Food and Agriculture Organization of the United Nations, Rome, 2006, http://www.virtualcentre.org/en/library/key_pub/longshad/A0701E00.htm.

18 Steinfeld et al., *Livestock's Long Shadow*.

19 S. Parente and E. Lewis-Brown, *Freshwater Use and Farm Animal Welfare*, CIWF/WSPA, 2012, http://www.ciwf.org.uk/includes/documents/cm_docs/2012/f/freshwater_use_and_farm_animal_welfare_12_page.pdf.

20 World Economic Forum, 'The bubble is close to bursting: A forecast of the main economic and geopolitical water issues likely to arise in the world during the next two decades', draft for discussion at the World Economic Forum Annual Meeting 2009.

21 Based on a bathtub holding 175 litres of water, and published figures showing the total water footprint for a kilo of beef, pork and chicken being 15,500 litres; from Water Footprint Network, http://waterfootprint.org/?page=files/Animal-products (accessed 10 September 2012).

22 *Wired* magazine, 'Peak Water: Aquifers and Rivers Are Running Dry. How Three Regions are Coping', 21 April 2008, http://.wired.com/science/planetearth/magazine/16-05/ff_peakwater?currentPage=all (accessed 10 September 2012); R. Courtland, 'News briefing, Enough water to go around?', *Nature*, published online 19 March 2008, doi:10.1038/news.2008.678, www.nature.com/news/2008/080319/full/news.2008.678.html.

23 C. S. Smith, 'Al Kharj Journal; Milk Flows from the Desert At a Unique Saudi Farm', *New York Times*, 31 December 2002, http://www.nytimes.com/2002/12/31/world/al-kharj-journal-milk-flows-from-the-desert-at-a-unique-saudi-farm.html (accessed 10 September 2012).

24 B. Bates et al., *Climate Change and Water*, IPCC Technical paper VI, IPCC, WMO and UNEP, 2008, http://www.ipcc.ch/pdf/technical-papers/climate-change-water-en.pdf.

25 J. Bruinsma, *The resource outlook to 2050: by how much do land, water and crop yields need to increase by 2050?*, Expert meeting on 'How to Feed the World in 2050', Rome, FAO, 24–26 June 2009.

26 BBC, 'Hundred-Dollar hamburger?' 14 June 2011, http://www.bbc.co.uk/news/business-13764242 (accessed 12 September 2012).

27 Government Office for Science, *Foresight Project on Global Food and Farming Futures. Synthesis Report C1: Trends in food demand and production*, January 2011; S. Msangi and M. W. Rosegrant, Agriculture in a dynamically-changing environment: IFPRI's long-term outlook for food and agriculture under additional demand and constraints, paper written in support of Expert Meeting on 'How to feed the World in 2050', Rome, FAO, 2009, http://www.fao.org.wsfs/forum2050/wsfs-back-ground-documents/wsfs-expert-papers/en/; H. Steinfeld et al., *Livestock's Long Shadow, environmental issues and options*, FAO, Rome, 2006, Introduction, p.12.

28 P. J. Gerber, H. Steinfeld, B. Henderson, A. Motter, C. Opio, J. Dijk-man, A. Falucci and G. Tempio, 'Tackling climate change through livestock – a global assessment of emissions and mitigation opportunities', Food and Agriculture Organisation of the United Nations (FAO), Rome, 2013.

29 Calculated from FAOSTAT online figures for global grain harvest (2009) and food value of cereals, based on a calorific intake of 2,500 kcalories per person per day.

30 Oxfam, *4-a-week: changing food consumption in the UK to benefit people and the planet*, Oxfam GB Briefing Paper, 2009.

31 *The Economist*, 'Food and the Arab Spring: Let them eat Baklava; Today's Policies are Recipe for Instability in the Middle East', 17 March 2012, http://www.economist.com/node/21550328; UN, 'Soaring cereal tab continues to affect poorest countries, UN agency warns', UN News Centre, 11 April 2008, www.un.org/apps/news/story.asp?NewsID=26289&Cr=food&Cr1=prices.

32 Defra, 'Farm scale evaluations: Managing GM crops with herbicides: effects on farmland wildlife', 2005.

33 M. A. Altieri, 'The Ecological Impacts of Large-Scale Agrofuel Monoculture Production Systems in the Americas', *Bulletin of Science Technology Society*, June 2009, 29(3), pp.236–44, http://bst.sagepub.com/content/29/3/236.

34 GMO Compass, 'Genetically modified plants: global cultivation area: Soybeans, Maize', 2010, www.gmo-compass.org/eng/agri_biotechnology/gmo_planting/342.genetically_modified_soybean_global_area_under_cultivation.htm; www.gmo-compass.org/eng/agri_biotechnology/gmo_planting/341.genetically_modified_maize_global_area_under_cultivation.html.

Center for Human Rights and Justice, *Every Thirty Minutes: Farmer Suicides, Human Rights, and the Agrarian Crisis in India*, New York, NYU School of Law, 2011.

Independent, 'Charles: "I blame GM crops for farmers' suicides"', 5 October 2008, http://www.independent.co.uk/environment/green-living/charles-i-blame-gm-crops-for-farmers-suicides-951807.html (accessed 12 September 2012).

D. L. Bartlett and J. B. Steel, 'Monsanto's Harvest of Fear', *Vanity Fair*, May 2008, http://www.vanityfair.com/poli'cs/features/2008/05/monsanto200805 (accessed 30 July 2012).

I. Potrykus, 'The "Golden Rice" tale', Agbioworld, http://agbioworld.org/biotech-info/topics/goldenrice/tale.html (accessed 30 July 2012).

New York Times, 'Scientist at work: Ingo Potrykus; Golden Rice in a Grenade-Proof Greenhouse', 21 November 2000/11/21/science/science-at-work-ingo-potrykus-golden-rice-in-a-grenade-proof-greenhouse.html?

http://agropedia.iitk.ac.in/?q=content/golden-rice (accessed 30 July 2012).

Golden Rice Project, Golden Rice Humanitarian Board website, http:// www.goldenrice.org/index/php (accessed 30 July 2012).

Yahoo voices, 'Dr Duane Carl Kraemer'; Hawthorne, 'A Project to Clone Companion Animals'; Warner, 'Inside the Very Strange World'.

H. Pickett, 'Farm Animal Cloning', Godalming, Compassion in World Farming, 2010, http://www.ciwf.org.uk/includes/documents/cm_docs/2010/c/compassion_2010_farm_animal_cloning_report.pdf.

P. Loi, L. della Salda, G. Ptak, J. A. Modliński and J. Karasiewicz, 'Peri-and post-natal mortality of somatic cell clones in sheep', *Animal Science Papers and Reports*, 22 (Suppl. 1) (2004, pp. 59–70.

EFSA, 'Scientific Opinion of the Scientific Committee on a request from the European Commission on food safety, animal health and welfare and environmental impact of animals derived from cloning by somatic cell nucleus transfer (SCNT) and their offspring and products obtained from those animals', *EFSA Journal, 767* (2008), pp. 1–49.

EGE, Opinion No. 23: 'Ethical Aspects of Animal Cloning for Food Supply', The European Group on Ethics in Science and New Technologies to the European Commission, 16 January 2008.

European Commission, 'Europeans' attitudes towards animal cloning', October 2008, http://ec.europa.eu/public_opinion/flash/fl_238_en.pdf.

Daily Telegraph, 'Genetically modified cows produce "human milk"', 2 April 2011, http://www.telegraph.co.uk/earth/agriculture/geneticmodification/8423536/Genetically-modified-cows-produce-human-milk.html# (accessed 13 September 2012).

Chapter 5 The Solutions

4 January 1986. In R. Body, *Our Food, Our Land*, Rider, London 1984.

M. Bittman, 'Don't End Agricultural Subsidies, Fix Them',
New York Times, 1 March 2011, http://opinionator.blogs.nytimes.com/2011/03/01dont-end-agricultural-subsidies-fix-them/; J. Steinhauer, 'Farm Subsidies Become Target Amid Spending Cuts', *New York Times*, 6 May 2011, http://www.nytimes.com/2011/05/07/us/politics/07farm.html

J. Bruinsma, *The resource outlook to 2050: By how much do land, water and crop yields need to increase by 2050?*, FAO Export Meeting on 'How to Feed the World in 2050', FAO, Rome, 24–26 June 2009; United Nations, *World Economics and Social Survey 2011: The great green technological transformation*, United Nations, New York, 2011.

Government Office for Science, *Foresight Project on Global Food and Farming Futures Synthesis Report C1: Trends in food demand and production, 2011; environment:IFPRI's long term outlook for food and agriculture under additional demand and constraints*, Expert Meeting on 'How to feed the World in 2050', Rome, FAO; H. Steinfeld et al., Livestock's Long Shadow, environmental issues and options, FAO, Rome, 2006, Introduction, p.12.

Calculated from FAOSTAT online figures for global grain harvest (2009) and food value of cereals. Based on a calorific intake of 2,500 kcalories per person per day.

Steinfeld et al., *Livestock's Long Shadow*, p. 43.

FAO, *State of the World Fisheries and Aquaculture*, 2010.

David Pimentel et al., 'Reducing energy inputs in the US food system', *Human Ecology*, 36 (2008), pp.459–71.

T. Stuart, *Waste: Uncovering the global food scandal*, Penguin 2009.

S. Fairlie, *Meat – a Benign Extravagance*, Permanent Publications, 2010, see pp. 46–50.

J. Gustavsson, C. Cederberg, U. Sonesson et al., *Global Food Losses and Food Waste: extent, causes and prevention*, FAO, Rome, 2011, www.fao.org/fileadmin/user_upload/ags/publications/GFL_web.pdf.

UN FAO, 2013, Food wastage footprint: impact on natural resources, http://www.fao.org/docrep/018/i3347e/i3347e.pdf (accessed 13th September 2013).

P. Stevenson, 'Feeding nine billion: How much extra do we need to produce?', 13 June 2013, http://www.eating-better.org/blog/3/Feeding-nine-billion-how-much-extra-food-do-we-need-to-produce.html (accessed 25 July 2013).

Stuart, *Waste*.

Lester R. *Brown, Plan B 4.0: Mobilising to save civilization*, Earth Policy Institute, W. W. Norton, 2009.

EEA, *The European environment – state and outlook 2010: synthesis*, European Environment Agency, Copenhagen, 2010.

Author's calculation.

B. White, *Alaska Salmon Fisheries Enhancement Program Report 2010*, Annual Report, Alaska Department of Fish and Game, 2011, http://www.adfg.alaska.gov/FedAidPDFs/FMR11-04.pdf (accessed 27 September 2012).

M. Kaeriyama, 'Hatchery Programmes and Stock Management of Salmonid Populations in Japan', in Howell et al., *Stock Enhancement and Sea Ranching*.

S. D. Sedgwick, *Salmon Farming Handbook*, Fishing News Books, Surrey, 1988.

INDEX

191

ACKNOWLEDGEMENTS

I would like to offer particular thanks to my co-author and collaborator Isabel Oakeshott, for crossing continents to help document the subject matter that we poured into this book. I am hugely grateful for her hard work, her instinct for storytelling, and for enduring some pretty awful sights during our travels. I owe a huge debt of thanks to our researcher, Jacky Turner, without whose thoroughness and commitment this book would have been much the poorer.

Grateful thanks go too to Tina Clark for her endless patience, help with the manuscript, planning field trips and providing essential support so reliably with unfaltering ease.

For this picture-led version of *Farmageddon*, I must give particular thanks to my crew in the field. Firstly, to Jim Wickens for his encyclopedic insight into the issues, of where to go and who to talk to, as well his keen eye for a picture. To our camera crews and support: Alejandro Reynoso, Brian Kelley and Jim Philpott. Many thanks to Luke Starr for research related to our field trips, and to Laurence Stephenson for help in editing pictures and video arising from the trip. To Veronica Oakeshott for documenting our trip through China, and to Jeff Zhou for seeing us through some difficult moments.

Many thanks to Charlotte Reid for enormous help with editing the original manuscript for this picture version. To Carol McKenna for overseeing the project with such enthusiasm, to Angela Wright, Natasha Boyland, Nicki Rust, Phil Brooke and Wendy Smith for additional research and to Sarah Bryan for sourcing and collating all the images.

Grateful thanks to Peter and Juliet Kindersley for inspiring and supporting this book; in seeing the potential and helping us make it a reality.

Special thanks to Jayne Parsons at Bloomsbury for believing in *Farmageddon in Pictures* and for making the book happen. And to Bill Swainson, commissioning editor for the original book at Bloomsbury on which this version is based. Thanks are due to Peter Bridgewater at Ivy Press, and to Jonty Whittleton, for coming up with the original concept layout and to Nick Clark for bringing the concept to life in a creative design. Grateful thanks to Valerie James, Jeremy Hayward, Mahi Klosterhalfen, Sir David Madden, Reverend Professor Michael Reiss, Michel Vandenbosch, Rosemary Marshall, Sarah Petrini and Teddy Bourne for steadfast belief and support throughout. Thank you too to Robin Jones, my literary agent, for support and encouragement.

Finally, huge thanks to my wife, Helen, for her understanding, particularly during long periods away from home so soon after our marriage.